Peter Lund Simmonds

Animal Products

Vol. I

Peter Lund Simmonds

Animal Products
Vol. I

ISBN/EAN: 9783337240028

Printed in Europe, USA, Canada, Australia, Japan

Cover: Foto ©berggeist007 / pixelio.de

More available books at **www.hansebooks.com**

SOUTH KENSINGTON MUSEUM SCIENCE HANDBOOKS.

[BRANCH MUSEUM, BETHNAL GREEN.]

ANIMAL PRODUCTS.

This work, prepared by order of the Lords of the Committee of Council on Education, is intended to serve, in the first instance, as a descriptive guide to the Collection of Animal Products at the Bethnal Green Branch of the South Kensington Museum; and, secondly, as a practical treatise on Economic Zoology for the use of the Public generally.

While there are very many treatises on Natural History, scientifically arranged, there are few works in which the uses of Animals to Man, and the important Commercial Products they furnish, have received prominent attention. The author has long had this extensive subject before him in cataloguing, labelling, and describing the large and increasing Collection of Animal Products in the Bethnal Green Museum, and believes that a detailed description of the various useful substances furnished by the Animal Kingdom may be rendered both instructive and interesting.

ANIMAL PRODUCTS

THEIR PREPARATION, COMMERCIAL USES, AND VALUE.

BY

P. L. SIMMONDS,

EDITOR OF THE "JOURNAL OF APPLIED SCIENCE."

Published for the Committee of Council on Education

BY

SCRIBNER, WELFORD, AND ARMSTRONG,

NEW YORK.

CONTENTS.

INTRODUCTION.

CHAPTER I.

THE WOOL-PRODUCING ANIMALS AND THEIR USES TO MAN.

CHAPTER II.

PRODUCTS OF WOOL-BEARING ANIMALS.

CHAPTER III.

PRODUCTS OF THE BOVINE TRIBE.

CHAPTER IV.

OTHER ECONOMIC PRODUCTS OF CATTLE.

CHAPTER V.

THE DEER AND ANTELOPE TRIBES.

CHAPTER VI.

FURS AND THE FUR TRADE.

CHAPTER VII.

CARNIVOROUS MAMMALS.

CHAPTER VIII.

CARNIVOROUS MAMMALS—(*continued*).

PAGE

CHAPTER IX.

CARNIVOROUS MAMMALS—(*concluded*).

CHAPTER X.

MAMMALS OF SECONDARY IMPORTANCE, THEIR ECONOMIC PRODUCTS.

CHAPTER XI.

SOLIDUNGULA AND PACHYDERMS AND THEIR USES.

CHAPTER XII.

PACHYDERMS, OR THICK-SKINNED ANIMALS, AND THEIR PRODUCTS.

CHAPTER XIII.

AQUATIC MAMMALS AND THEIR PRODUCTS.

CHAPTER XIV.

MARSUPIALS AND THEIR USES.

LIST OF ILLUSTRATIONS.

ANIMAL PRODUCTS.

ORIGIN OF THE ANIMAL PRODUCTS COLLECTION.

At the close of the Great Exhibition of 1851, many of the articles there displayed were presented to Her Majesty's Commissioners by various Foreign Governments, and numerous individual exhibitors, to form the nucleus of a permanent Trade Collection. It was considered that such a Collection would not only be interesting as constituting a lasting memorial of the Exhibition and a record of the state of Industry in 1851, but that it might be rendered of great practical benefit to the Manufacturing and Mercantile communities if systematically arranged for purposes of reference, with a view both to technical instruction and to the ever-changing and increasing wants of trade in this great commercial country. The Collection thus presented to the Commissioners contained many and valuable specimens in the three great kingdoms of Animal, Vegetable, and Mineral products. Great progress had been made in the development of two important National Collections illustrative of the Vegetable and Mineral kingdoms respectively, viz., the Museum of Economic Botany at Kew, and the Museum of Practical Geology in Jermyn Street, and to these the vegetable and mineral products were respectively consigned.

No corresponding Collection for the Animal Kingdom had, however, previously existed. The Royal Commissioners, therefore, thought it desirable to endeavour to supply this deficiency by the formation of a *Collection of Animal Products*, the articles of that class, presented to them in 1851, serving as an appropriate nucleus for such a Collection. The Society of Arts, being equally impressed with the importance of this object, co-operated with the Commissioners towards its attainment, and joined in securing the services of Professor Solly for a period of two years, ending in 1855, to superintend the formation of the Collection; Dr. Lyon Playfair, M.P., the then scientific referee of the Department of Science and Art, giving valuable assistance in the development and arrangement of the articles, a work in which the author of these pages also took a not inconsiderable part. The Collection was first exhibited to the public in 1857; and in 1858 the whole of the Collection of Animal Products, as it then existed, was presented by Her Majesty's Commissioners for the Exhibition of 1851 to the Science and Art Department of the Committee of Council on Education. It was then exhibited as part of the South Kensington Museum, where it remained for several years, but was subsequently removed to the Branch Museum of the Department at Bethnal Green, where it is now arranged in the *Lower Gallery* on the *South Side* of the Building.

As the Food Products from Animals, and the Economic relations of Insects to man, are dealt with in separate Handbooks or Guides to those Collections by other writers, the various products of the Mammalia alone are proposed to be touched upon in this Manual, which will be devoted chiefly to those used for Manufactures, without following in strict order the natural history classification; but tracing the Collection as nearly as possible in the methodical manner in which it has been arranged, and adding to the text such illustrations of animals and their products, as may serve to render it interesting and useful, and help to make it more generally understood.

THE CLASSIFICATION (beginning at the *East* end of the Gallery) adopted originally in the arrangement of the COLLECTION of ANIMAL PRODUCTS is as follows:—

CLASS I.—*Animal Substances employed for Textile Manufactures and Clothing.*

Division I. Wool, Mohair, and Alpaca.
,, II. Hair, Bristles, and Whalebone.
,, III. Silk.
,, IV. Furs.
,, V. Feathers, Down, and Quills.
,, VI. Gelatin, Skins, and Leathers.

CLASS II.—*Animal Substances used for Domestic and Ornamental Purposes.*

Division I. Bone and Ivory.
,, II. Horns and Hoofs.
,, III. Tortoise-shell.
,, IV. Shells and Marine Animal Products for Manufacture, Ornament, &c.
,, V. Animal Oils and Fats.

CLASS III.—*Pigments and Dyes yielded by Animals.*

Division I. Cochineal and Kermes.
,, II. Lac and its applications.
,, III. Nut Galls, Gall Dyes, Blood, &c.
,, IV. Sepia, Tyrian Purple, Purree, &c.

CLASS IV.—*Animal Substances used in Pharmacy and in Perfumery.*

Division I. Musk, Civet, Castoreum, Hyraceum, and Ambergris.
,, II. Cantharides, Leeches, &c.

CLASS V.—*Application of Waste Matters.*

Division I. Guts and Bladders.
,, II. Albumen, Casein, &c.
,, III. Prussiates of Potash and Chemical Products of Bone, &c.
,, IV. Animal Manures.—Guano, Coprolites, Animal Carcases, Bones, Fish Manures, &c.

The special object of this Collection is not merely the formation of a Museum showing the various Animal Products entering into British and Foreign Commerce, but, at the same time, to instruct and inform the visitor as to the magnitude of the trade, the varieties, peculiar characteristics and suitability for various purposes, of different substances. While, therefore, the mere visitor for pleasure will be gratified by a passing glance at such a general collection of useful and ornamental products, the more thoughtful and inquiring will here find ample opportunities presented to them of studying quietly, systematically, and in progressive detail, the principal Arts and Manufactures arising out of Animal substances which result in such individual benefit, and contribute so greatly to our national wealth and extensive commerce.

The objects exhibited are arranged into classes, groups, and subdivisions, which proceed step by step from the raw material, through the various stages of manufacture, up to the finished product.

The Food Products of Animal origin are illustrated in the Food Collection arranged in the opposite gallery.

Descriptive general, and special, framed labels are hung about the galleries; every case, article, and particular manufacture is fully described, and cheap catalogues are also on sale, so that the visitor will have little difficulty in gleaning useful information as he proceeds.

The extent and importance of the trade in Animal Products generally is probably little understood. It would be indeed difficult to form a precise estimate of its magnitude even as regards the trade, industry, and money value for the United Kingdom alone, but there are some few data from which an approximate account of the raw materials and manufactures derived from the Animal Kingdom may be formed. We have tolerably correct agricultural statistics of our domestic Live Stock, and also official records made annually by the Board of Trade, of the value of the Imports and Exports. If we take, therefore, the data from these for the year 1875 (which, it may be incidentally remarked, was not a very prosperous trade year), we shall arrive at some idea of the enormous figures and the large interests involved.

Value of the *Imports* of Animal origin brought into the United Kingdom in 1875.

Live Animals	£8,466,226
Food Products:	
Bacon and Hams	6,982,470
Pork, salted and fresh	590,356
Beef, salted	357,201
,, fresh or preserved	97,136

Butter	8,502,084
Cheese	4,709,508
Eggs	2,559,860
Fish, fresh	218,031
,, cured or salted	1,048,546
Poultry and Game, including Rabbits	328,044

Articles chiefly for Manufactures:

Bones for manufacturing	£74,059
Bristles	419,203
Cochineal	492,976
Feathers for beds	126,177
,, for ornament	713,199
Galls	63,359
Lac resin and lac dyes	806,117
Hair of various kinds	1,483,984
,, manufactures	68,323
Hats of felt	51,498
Hides, wet and dry	4,203,371
,, tanned or otherwise prepared	2,814,042
Horns and Hoofs	172,966
Isinglass	86,443
Lard	1,634,769
Leather manufactures	308,290
,, Boots and Shoes	240,000
,, Gloves	2,430,876
Manures: Bones of Animals	630,656
,, Guano	1,293,436
Oils: Train or Blubber	489,817
,, Spermaceti or head matter	427,884
,, Animal	37,433
Rags, Woollen, for re-working	599,402
Silk, raw and thrown	3,546,456
,, waste	415,085
,, manufactures of all kinds	12,264,532
Skins of various kinds (not furs)	2,494,979
Furs, Pelts, and manufactures of Furs	1,375,512
Specimens of Natural History	22,785
Tallow	2,045,863
Ivory	772,371
Wax	118,549
Whale-fins	42,240
Wool of various kinds	23,451,887
Woollen yarn	1,491,117
,, manufactures	4,308,357
	£105,577,155

These are only the principal articles; sponges, mother-of-pearl, cowries, and other shells, tortoise-shell, leeches, cantharides, and many other products, are not included, being aggregately grouped by the Board of Trade under "Miscellaneous Articles."

There are also certain vegetable substances imported, which are chiefly used in the preparation and manufacture of Animal Products, such as all the tanning materials, which should be taken into consideration; these include the following values:—

Barks and Extracts for tanning	£395,318
Cutch	140,150
Gambier	601,105
Sumach	246,343
Valonia	622,019
Myrobalans	100,000
	£2,104,935

Value of the Animals and Animal Substances, the produce of the United Kingdom, *exported* in 1875.

Food substances:	
Butter	£240,281
Cheese	88,143
Fish	1,192,481
Provisions	693,294
Horses	241,106
Manufactures and materials, &c.:	
Hats of all sorts	1,045,440
Leather manufactures	3,881,168
Silk, Thrown, Twist and Yarn	880,923
Silk manufactures	1,734,519
Skins and Furs of all sorts	946,694
Soap	310,511
Umbrellas and Parasols	356,467
Wool	928,264
Woollen and Worsted Yarn	5,099,307
Woollen manufactures	21,659,325
	£39,297,923

This is exclusive of a number of minor articles not enumerated or specified in the Board of Trade Returns.

Taking the latest Agricultural Returns as our guide for numbers, and assuming a very moderate value for each animal, we get at the following approximate estimate of the value of our domestic stock.

Live Stock in Great Britain and Channel Islands, 1875 :

Horses*	1,349,691	at	£16	0	£21,587,056
Cattle	6,050,797	,,	10	0	60,507,970
Sheep	29,243,790	,,	1	10	43,865,685
Swine	2,245,932	,,	1	5	2,807,415
					£128,768,126

There are no returns of asses and mules, goats and poultry, for Great Britain.

Number and value of Live Stock in Ireland in 1875:†

Horses and Mules . .	547,675	at	£10	0	£5,476,750
Asses	179,742	,,	1	0	179,742
Cattle	4,111,990	,,	10	0	41,119,900
Sheep	4,248,158	,,	1	10	6,372,237
Goats	268,894	,,	0	10	134,447
Pigs	1,249,235	,,	1	5	1,561,544
Poultry	12,055,768	,,	0	1	602,788
					£55,447,408

When we find that the figures we have quoted give an estimated money value exceeding £331,000,000 sterling, and that to this has to be added all the dairy produce, the poultry and their products for Great Britain; the annual clip of British wool, which may be estimated at 160,000,000 lbs., worth at least £8,000,000; the hides and skins, tallow, horns, bones, and other offal, horse and cow hair, woollen rags collected, the game and rabbits, the sea and river fisheries; besides the products of our woollen, leather, glove, silk, soap and comb manufactures retained for home consumption, furs, brushes, and many other articles, we ought to add a great many millions more to the aggregate value or total.

* Returned by occupiers of land alone, and quite exclusive of horses kept in towns, racehorses, &c., of the numbers of which there are no complete returns.

† These values are the old official prices, and many are much below the present value for horses, &c.

These collective figures of the value of our Live Stock and of the Animal Products imported and exported, will at least show what a large amount of capital is invested in them, and that they must necessarily give busy and remunerative occupation to a great number of persons in the raising, collection, distribution and after preparation of most of the articles, to fit them for various uses, whilst the amount of shipping tonnage employed, and the inland transport by road and rail from place to place, of the raw materials and the finished manufactures, are other great sources of active industry, in which numbers of our population are specially interested.

INTRODUCTION.

CLASSIFICATION OF MAMMALS.

NATURALISTS have described more than 2,000 species of Mammals, or animals which suckle their young. They are characterised by warm red blood, and breathe by means of lungs. They have been grouped into different Orders, each divided into Genera, which usually include several individual Species.

Amongst them are the cattle of our fields, beasts of burden, and domesticated animals of many kinds. Most of these are familiar to all; but a more perfect knowledge of their nature contributes to the improvement of agricultural stock, affords indications of rational methods of treating the diseases to which they are subject, and makes us acquainted with the sources of supply of many Animal Products largely used for food or in the various arts and manufactures.

The following, although it may not satisfy all as a classification of Mammals, is yet sufficiently clear for the purposes of this work, which is intended as a description of the Economic uses of the Animals, and their Commercial products, rather than a treatise on Systematic Zoology.

I. QUADRUMANA, or four handed.* *Examples*—Ape, Baboon, Monkey.

There are about 100 species belonging to this order at present known. In some countries the flesh of monkeys is eaten. The skins and skeletons form articles of commerce, and live animals are purchased for zoological gardens.

* The order Bimana—Man—has been passed over, the only products of any commercial value derived from the human race being the hair of females, in which some considerable trade is carried on, and skulls and skeletons for museums. In the Waste Products Collection will be found illustrations of the use of human hair, and in Case **87** is a piece of cloth made with human hair; ladies' muffs have also been made of it.

II. CHEIROPTERA. *Example*—Bats.

Some of these are fruit eaters, others insect eaters. The flesh of a few bats is eaten.

III. INSECTIVORA. Insect eaters. *Examples*—The Hedgehog and Mole.

This order is not of much commercial importance.

IV. CARNIVORA. Flesh eaters—beasts of prey. *Examples*—Lion, Cat, Fox, Bear, Seal.

This is the most important order of animals for the supply of skins and furs to commerce. The flesh of some carnivorous animals is eaten in certain districts.

V. RODENTIA. Gnawing animals. *Examples*—Rat, Hare, Rabbit, Beaver, Squirrel.

Many furnish furs to commerce, and their flesh serves for food. 380 species are known to naturalists.

The great incisor teeth being separated widely from the molars is the characteristic of this order, and enables the animals to gnaw hard substances, such as wood, with facility.

VI. EDENTATA. Animals wholly or partially without teeth. To this order are added the Monotremes. *Examples*--Armadillo, Sloth.

This order is commercially unimportant.

VII. RUMINANTIA. Ruminants, or cud chewers, who remasticate their food, which, after it is brought into the first stomach and imperfectly digested, comes again to the mouth.

They are cloven-hoofed, and with but few exceptions have horns. The horns are either solid, deciduous, as in the deer, persistent, with a core, as in the antelope and goat, or round and smooth as in the ox and buffalo: some of the buffaloes, however, have them wrinkled. *Examples*—Ox, Sheep, Goat, Camel, Alpaca.

This is the most valuable order of any for the commercial products it supplies, in animal food, skins, wool, tallow, &c.

VIII. SOLIDUNGULA, or solid hoofed, on each foot only one toe or hoof. *Examples*—Horse, Ass.

These are exceedingly useful to man as draught animals and for r products, hair, skins, flesh, &c.

IX. PACHYDERMATA. Thick-skinned animals. *Examples*—Elephant, Hippopotamus, Hog.

By some naturalists the Pachydermata are made to include the Solidungula.*

Some of the animals of this order, as the hog and the elephant, are of high commercial importance.

X. AQUATIC MAMMALS. *Examples*—Manatee, Dugong, Beluga, Whale.

By naturalists the manatees are grouped in a separate order termed Sirenia, the seals and otters belong to the Carnivorous order, and the whales to the Cetacea.

The aquatic mammals are commercially valuable for their flesh as food, for their skins, and the oil obtained from their blubber.

XI. MARSUPIALIA. Pouched animals. *Examples*—Kangaroo, Opossum, Wombat.

The animals of this order are sought for their flesh and skins; some are herbivorous, others carnivorous.

* One of the most recent classifications is that establishing the order UNGULATA, hoofed mammals, which combines the three divisions, Pachydermata, Solidungula, and Ruminantia, which have been thrown into three new sections or sub-orders, and stand classified thus, according to Mr. Wilson (Elements of Zoology):—

A. ARTIODACTYLA.	B. PERISSODACTYLA.	C. PROBOSCIDEA.
§ 1. Omnivora.	§ 1. Solidungula.	Elephants.
§ 2. Ruminantia.	§ 2. Multungula.	

Mr. ANDREW MURRAY classifies the UNGULATA in the following manner:—

A. MONODACTYLA (The *Solidungula*). Horses, &c.

B. ARTIODACTYLA.

1. Ruminants, including camels, oxen, sheep, antelopes, camelopard, deer, musk deer and chevrolins (Tragulidae).
2. Anoplotheridae (extinct).
3. Non Ruminants, including the peccary, swine, hippopotamus.

C. MULTUNGULA.

Tapiridae—tapirs.

Nasicornia—rhinoceros.

CHAPTER I.

THE WOOL-PRODUCING ANIMALS AND THEIR USES TO MAN.

The Collection commences at the east end of the raised floor gallery, immediately facing the main entrance. It starts with the woolly coverings of animals in all their variety, and illustrates the economic uses to which these are applied—for clothing for the human race, for fabrics of different kinds, carpets, felts, &c. This preliminary Chapter is specially devoted to a description of the varieties of the Sheep; the different breeds or races which careful culture has produced; the characteristics of the kinds of wool obtained from special breeds in various countries. The average weights of the clip of wool from the fleeces of different sheep are given, official statistics of the number of sheep in different countries, the wool produce of the world, and the sources of our foreign supply of wool. The classifications adopted in sorting out fleeces, the special distinctive characters of wool, fur, and hair are pointed out, and their chemical composition; the processes of shearing, wool-washing, scouring, and dyeing, are then touched upon.

THE SHEEP.—Of the domesticated animals the Ruminants among the Mammals are the most serviceable to man, and have multiplied and been diffused more generally over the face of the globe than any others. Their commercial products are also of the greatest importance, and as the Ovine race stand, perhaps, the highest in estimation for their direct use, we commence with a description of those of the Sheep, as the principal wool-producing animal.

Of the original breed of this invaluable animal, nothing certain is known. Several varieties of wild sheep have by naturalists been considered entitled to the distinction of being the parent stock, and the marked differences between the wild and domestic species are readily accounted for by the known variability of the animal. No other animal seems to yield so submissively to the manipulations of culture.

The sheep gives immediate employment to thousands, who in their several spheres utilise different parts of it for the various uses of the great human family. Among these we have the breeder, the butcher, the skinner, the tanner, shoemaker, tallow chandler, etc. Then the "fleece," which we call wool, gives occupation to the wool-brokers, wool-staplers, spinners, manufacturers, clothiers, and many subordinate branches of trade to which these give rise.

On account of its numerous useful properties, the sheep has deservedly become an object of national consideration in almost all temperate countries.

It is of the most extensive utility to man. We are clothed by its fleece; the flesh is a delicate and wholesome food; the skin dressed forms different parts of our apparel, and is used for various economic purposes. The entrails, properly prepared and twisted, serve as strings for musical instruments. The calcined bones have industrial uses. Sheep's milk is thicker than that of cows, and consequently yields a greater quantity of butter and cheese. From the Larsac race of sheep in France, the celebrated Roquefort cheese is made to the extent of about 6 or 7 million pounds annually. There is no manure so fertilising as that of the sheep, and it does not so readily waste by exposure as that of other animals. A German agriculturist has calculated that the droppings from 1000 sheep during a single night would manure an acre of land sufficiently.

If we look next at sheep as a source of our animal food supply, having regard only to the United Kingdom, we find that the agricultural returns of 1874 gave the number of sheep and lambs at 34,800,000. Now it is estimated by good authorities that half of our stock of sheep are slaughtered annually, and as these 17,400,000 animals will average 56 lbs. per head, we have thus an annual supply of 8,700,000 cwts. of meat per annum, besides 1,000,000 imported animals, which will give about 450,000 cwts.

more of mutton. The average wholesale price of mutton per stone of 8 lbs. in the Metropolitan market is now 6*s.* 5*d.* against 4*s.* 5*d.* a quarter of a century ago. When we consider their value, also, for food, on the Continent, in America and the Colonies, and the quantity of tallow they yield, as well as skins for the tanner and glove maker, we shall begin to understand how immense is the value of the Ovine race to man, both for sustenance and clothing. In many foreign countries, the flesh of the sheep is disliked, or at least rarely eaten, and the animal is tended solely for its fleece. In Spain, except by the poorest, mutton is considered unfit for food.

The following figures give the number of sheep in various countries according to the latest official returns :—

EUROPE.		
Russia . . .	1870,	48,132,000
Sweden . .	1873,	1,695,434
Norway . . .	1865,	1,710,000
Denmark . .	1871,	1,842,481
Iceland . . .	1866,	800,000
German Empire .	1873,	24,999,406
Holland . .	1873,	901,515
Belgium . . .	1866,	586,097
France . .	1872,	24,589,647
Portugal . .	1870,	2,706,777
Spain . . .	1865,	22,054,967
Italy . . .	1874,	6,977,104
Austria Proper .	1871,	5,026,398
Hungary . .	1870,	14,289,130
Switzerland .	1866,	447,001
Greece . . .	1867,	2,539,538
Turkey . . .	1870,	16,000,000
Moldavia and Wallachia .	1873,	4,786,294
Great Britain . .	1876,	28,178,950
Ireland . .	1875,	4,248,158

AFRICA.		
Egypt . . .	1871,	184,899
Algeria . .	1866,	10,000,000
Cape Colony . .	1875,	11,008,339
Natal . . .	1873,	343,763

AMERICA.		
United States . .	1875,	33,783,600
British America .	1871,	3,337,763
Uruguay . .	1872,	20,000,000
Argentine Confederation . .	1875,	70,000,000
Falkland Islands	1875,	60,000

ASIA.		
Ceylon . .	1874,	61,453
Mauritius . .	1875,	28,036

AUSTRALASIA.		
New South Wales	1875,	22,872,882
Queensland . .	1874,	7,268,946
Victoria . .	1875,	11,221,036
South Australia .	1875,	6,120,211
Western Australia	1875,	748,536
Tasmania . .	1875,	1,714,168
New Zealand .	1874,	11,674,863

WEST INDIES.		
Jamaica . .	1869,	21,761
Martinique and Guadaloupe .	1865,	23,607

Although upwards of 160,000,000 pounds of wool are produced annually in the United Kingdom, yet it may almost be said that sheep are kept in this country, not so much for their fleece as for the meat their carcases furnish, and for the great benefit they confer on agriculture. They have indeed been aptly designated "the sheet anchor of British agriculture," so indispensable are they to the scientific farmers of the present day. But in improving the carcase of these valuable animals, great care has been bestowed upon their wool-producing capabilities, in order to preserve and increase as much as possible the quantity and quality of this important article. The wools of this country are therefore abundant in quantity, and of good, strong, and very useful qualities; although not of such fine description as the Saxony, Spanish, and other Merinoes.

The Animal Products Collection of the Bethnal Green Branch Museum is particularly rich in fine samples of fleeces, various kinds of wools, and stuffed heads of the principal breeds of sheep.

On the east wall will be noticed—

No. **1**. A fine skin and head of a Hampshire Down ram, with a fleece of two years' growth.

Nos. **2**, **3**, **4**. Stuffed specimens of small sheep from the Shetland Islands.

Nos. **7** and **8**. Two framed lithographs of South Down and Highland sheep.

No. **9**. An oil painting of one of the earliest improved Merino rams, known as the First Consul.

In two large glazed cases against the east wall are sixty fleeces, with displayed locks, of all the choicest wools, British and foreign.

A breed of sheep to produce fine wool is distinct from a breed to produce mutton and wool. Of fine-woolled sheep the pure Merino takes the first place, producing a heavier fleece and of equal quality. It has also another great advantage over the Saxon Merino, in being much larger, hardier, and less liable to disease.

The Saxon Merino is found by farmers, even in Saxony, to be so unprofitable, that the numbers kept are being gradually diminished.

WOOL PRODUCE OF THE WORLD.—The following table, compiled from the most reliable information obtainable, furnishes an approximate estimate of the production of sheep's wool in the principal countries of the world in 1874.

	POUNDS.
South America	350,000,000
Asia, including Russia in Asia	320,000,000
Australia	250,000,000
United Kingdom	200,000,000
United States	200,000,000
France	150,000,000
European Russia	150,000,000
Turkey, European and Asiatic	140,000,000
Spain	62,000,000
Austria	60,000,000
Germany, Netherlands, and Belgium	60,000,000
Hungary	45,000,000
North Africa	50,000,000
South Africa	50,000,000
Italy	40,000,000
Persia	40,000,000
Portugal	12,000,000
Canadian Dominion	12,000,000
Sweden and Norway	12,000,000
Denmark	10,000,000
Greece	10,000,000
Mexico	500,000
Total	2,223,500,000

STATISTICS OF OUR WOOL IMPORTS.—The demands of our Woollen Manufactures require a large and increasing supply of the raw material; of this only a small proportion, estimated at 160,000,000 to 200,000,000 lbs. annually, is produced in the United Kingdom, and hence we are largely dependent on foreign supplies. Fortunately, the British possessions are increasing their wool production year by year, and rendering us more independent of supplies from foreign countries, as the following figures of our imports will show:—

Year.	From Australia.	South Africa.	India.	Total Imports from all Countries.
	lbs.	lbs.	lbs.	lbs.
1831	2,493,000	48,000	———	31,652,000
1841	12,399,000	1,080,000	3,009,000	56,180,000
1851	41,800,000	5,817,000	4,550,000	83,311,000
1861	68,506,000	18,676,000	19,161,000	144,067,000
1871	182,710,000	32,972,000	18,153,000	319,385,000
1875	238,631,716	44,112,213	22,680,126	360,903,270

BRITISH BREEDS.—England has for a long period been cele-

TWO-YEAR OLD SOUTHDOWN, SHOWN AT TAUNTON, 1875.

brated for her sheep. The leading and most improved short-woolled breed is the small brown-faced Southdown, chiefly occupying the hills of Sussex, but also now diffused over other counties. Its fleece, short and fine, weighs from three to fou

DORSET RAM, SHOWN AT CROYDON, 1875.

pounds; and its mutton, fine in flavour and grain, weighs, in two-year-old wethers, about eighteen pounds the quarter.

COTSWOLD, SHOWN AT CROYDON AND TAUNTON, 1875.

The Dorset is another short-woolled sheep, see preceding page.

The only other breed now looked upon as pure stock are the white-faced, long-woolled sheep, which are known as Leicesters from their home being the great midland grass district, of which Leicester is the centre.* The numerous other breeds are generally traceable to some amalgamation by crossing and re-crossing.

The breeds of sheep in the British Islands may be divided into *two* principal groups:—1. Those of the plains, grass lands, and arable districts, which vary greatly in size, quality of mutton, and the weight of wool they produce; and, 2. The hill or mountain sheep, which are less variable in general character, although they are greatly altered by the quality of their native ground and the altitude at which they generally range. The bulk of the mutton and wool of the country is now produced by the flocks kept on the old pastures and such lands as are worth cultivation by the means applied to modern agriculture.

It is not necessary to enumerate and describe the numerous varieties and sub-varieties of breeds of sheep common in different parts of England and other countries. Sheep of the Cotswold long-wool breed have been known to reach the enormous weight of eighty-four pounds per quarter, or 336 lbs. the carcase; and fleeces weighing above twenty pounds are not uncommon.†

AVERAGE WEIGHT OF FLEECES.—In 1851 Mr. Thomas Southey, after extensive inquiries, took the average weight for the United Kingdom at 5 lbs. Considerable changes have taken place in the actual weights of fleeces, owing to improved breeding; and even during the last quarter of a century this has been the case with sheep bred in agricultural districts, though not so much with those bred on pasture lands. The weights, moreover, are

* Frame No. **29** shows the characteristics of this wool, and in Case **46** are samples and locks of long wool from Earl Fitzwilliam's flocks.

† For these four representations of prize sheep of special English breeds we are indebted to the courtesy of the proprietor of the *Agricultural Gazette.*

considered to vary from year to year as much as from a quarter to half a pound per fleece, according to the seasons and breed.

The average weight of the clip of half-breds is from 5½ to 7 lbs., of Leicesters 7 to 8 lbs. Some of the large breeds in Gloucester and Somerset will weigh 7 lbs., and in Devon and Cornwall, unwashed fleeces 7½ lbs. In the East Riding of Yorkshire a large breed with deep staple and bright hair weighs 8½ lbs. The average weight for Wales and Scotland is 4¾ lbs. The Irish fleece ranges from 6 to 6½ lbs.

Allowance should be made, in all wools unwashed or in the grease, of one third in weight for clean wool. The quantity as well as the quality of the wool yielded by the sheep varies much with the breed, the climate, the food, and consequently with the soil on which the food is grown. The Hereford sheep, which are kept lean, and give the finest wool, yield only 1½ to 2 lbs. of washed wool, while a Merino will often give a fleece weighing three times that amount.

The following figures give the number of sheep officially returned in the United Kingdom and principal British Colonies for the year 1874.

Great Britain	30,313,941
Ireland	4,411,698
Total for the United Kingdom	34,725,639
New South Wales	22,872,882
Victoria	11,221,036
South Australia	6,120,211
Western Australia	777,861
Queensland	7,268,946
Tasmania	1,714,168
New Zealand	11,674,863
Canadian Dominion	3,155,509
Prince Edward Island	147,364
Cape of Good Hope	12,000,000
Natal	343,763
Total for the principal British Colonies	77,296,603

As regards the number of sheep, Russia exceeds the United Kingdom proper, having 48,000,000 head; the United States approximate closely to Great Britain in the number; whilst the other principal European pastoral countries stand as follows, in round figures:—France, 24,600,000; Spain, 22,000,000; Prussia, 19,600,000; Hungary, 14,000,000; and Italy 7,000,000.

LORD CHESHAM'S SHROPSHIRE, SHOWN AT CROYDON, 1875.

CHARACTERISTICS OF WOOL.—Wool resembles hair in many of its peculiarities; the chief point of difference being, that while the surface of the latter is smooth, that of the former is imbricated, a quality upon which the felting power of wool depends. This difference is not, however, perceptible to the naked eye or touch; indeed it would be very difficult to point out any perceptible qualities distinctive of the two substances. The bristles of the hog and the fine wool of the lamb can readily be distinguished from one another, for these are the extreme examples of the two substances; but in many cases hair and wool pass so completely into each other, that it is often impossible to mark the line of demarcation; and they have the same chemical composition.

The importance of the industrial employments of wool can scarcely be overrated. The pelage of mammiferous animals is composed of various kinds of hair, the most important of which are silky hairs and woolly hairs, and according to various circumstances the one or the other of these varieties predominates. The physiological conditions of age, sex, food, and climate serve to vary the quality of the fleeces. It is therefore well to study the characters of the hairs employed in manufactures, especially the woolly ones, by ascertaining their length, diameter, elasticity, etc., under the microscope. Thus examined we shall find that wool presents fine transverse or oblique lines (an inch containing from 2,000 to 4,000), which indicate an imbricated scaly surface. This characteristic, and the twisted form of fine wool, are the qualities which make it valuable for manufactures.

It has been well observed by Dr. Crisp that the coverings of animals are wonderfully adapted to the climates and elements which they inhabit. Thus we have a warm thick fur in the extreme northern zone, a woolly coat for the sheep in the temperate regions, a thin hairy covering in the tropics—in the air the light and beautiful feather—in the water the crust or scale. What armourer could make a coat of mail to equal that in which the armadillo is invested? how well it protects the animal from the weapons of its assailants.

The common impression is that wool is confined to the sheep, but experience shows that a great variety of other animals produce it also, and that under the long hair of the goat, for example, there will generally be found a certain amount of true wool; and we might go even further, and consider that, with very few exceptions, the external covering of all mammalian animals is a variable mixture of hair and wool.

Of the two extreme contrasts of hair produced on domesticated animals, one is rigid, shining, coarse, well fixed. It is the "jarre" which exists nearly alone in the ordinary conditions of the horse and the ox. The other hair or down, hidden under the first, is distinguished by being more curled and tangled, and more dull than the jarre, and also much finer. It is the "wool" which exists nearly alone in the Merino sheep and in the Cashmere goats. Wool is greatly preferred to the jarre by the manufacturer, because it is much finer, curls more readily, and is found bristling with little scaly asperities (due to its mode of development) which render it more adapted for felting and the manufacture of tissues. From the coats of sheep, goats, rabbits, etc., the coarse and rigid hair is carefully removed.

HUMAN HAIR. RABBIT FUR.

Wool, we find, is not then peculiar to the sheep, but forms an undercoat beneath the long hair in very many animals. Articles for clothing have been made from the wool of the musk-ox of North America, and from the wool of the ibex of Little Thibet; but in these and other such instances, they have been produced as objects of curiosity rather than for any commercial purpose.

In the sheep, judicious management has in the course of years increased the growth of wool, and rendered the occurrence of hair unusual. Wherever attention has been paid to sheep-breeding, there a marked improvement has been manifested in the particular direction in which the improvement has been sought, whether in the carcase or in the fleece. The sheep produces the finest quality of wool in the warmer temperate and sub-tropical zones only.

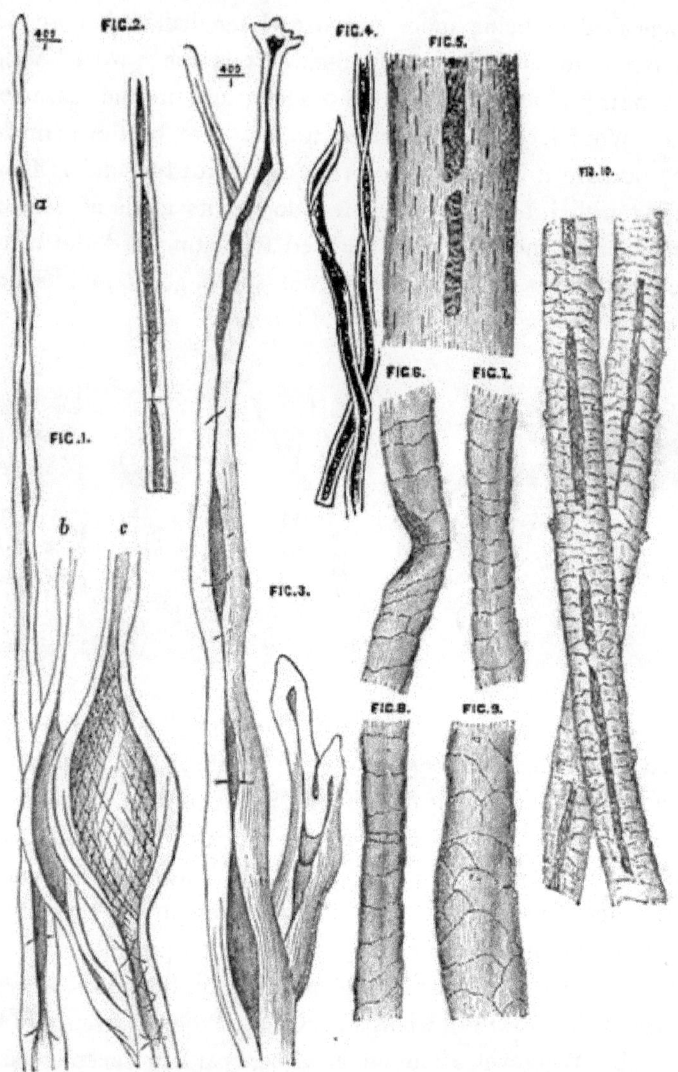

FIBRES OF FLAX, HEMP, JUTE, COTTON, AND WOOL MAGNIFIED.

Wool seems to be the only substance provided by Nature to satisfy all conditions required for beauty and utility in clothing the inhabitants of climates where extremes of heat and cold prevail. There is not a single property desirable in a fabric for human use that is not found in wool.

The engraving on the opposite page shows the contrast between vegetable and animal fibres as seen under the microscope.

Figures 1, 2, 3, 4 represent flax, jute, hemp, and cotton, and figures 5 to 10 magnified representations of coarse long, fine Saxony, and fine English wool, illustrating the difference of appearance.

The chemical composition of wool is as follows:

Carbon	50·65
Hydrogen	7·03
Nitrogen	17·71
Oxygen and sulphur	24·61
	100·

M. Chevreul, after analysing Merino wool heated to dryness above 100°, found it to consist of

Earthy matter left in water with which it was washed	26·06
"Suint" or fat, soluble in cold water	32·74
Other kinds of fats	8·57
Earthy matters collected with the fat	1·40
Pure wool	31·23
	100·000

Sulphur is a very important element in the composition of wool, and some close statistical calculations made, show that in the United Kingdom as much as five millions of pounds of sulphur are annually abstracted from the soil by the sheep. It is evident, therefore, that in order to have healthy animals and a full produce of wool, there must be in the soil a good supply of sulphur, nitrogen, potash, and phosphorus, or the pasture, combined with atmospheric influences, will not enable the animal to secrete wool in perfection.

Foreign Breeds of Sheep.—Attention to the cultivation of fine wool has long been paid in many countries, and has produced the highly valued Merino breed. It has been supposed that the female has more influence than the male on the bodily form of an animal; but the male, in sheep particularly, has been found to give the peculiar character to the fleece. The produce of a breed from a coarse-woolled ewe and fine-woolled ram will yield a fleece differing only one-fourth from that of the sire. By proceeding in the opposite direction, the wool would rapidly degenerate into its primitive coarseness. Great care must therefore be taken to exclude from a breeding flock any accidental varieties of coarse-woolled rams.

Immense services were rendered by the introduction of the Spanish sheep into England,* and in former times the sheep of Northern Africa into Spain. This excellence of the Merino consists in the unexampled fineness and felting property of its wool, which in fineness and in the number of imbrications and curves exceeds that of any other sheep the world produces.

SPANISH MERINO WOOL.

The thorough bred Negretti sheep takes a high and important position among the Merino breed. It has a deep built barrel-like body, powerful short legs, and compactly grown wool on the body and legs.

As wool is the principal object in Germany, the wethers are kept to ten and even thirteen years old, and they are then turned into inferior mutton. The sheep are housed in stables every

* In Case **61** are samples of the early products of Merino wool and yarn in this country in 1812 and 1829.

night, and carefully guarded in the daytime from the rain. Shearing takes place in May, and is done by women, who shear in a shearing house with strong scissors. Washing is very carefully attended to, and is done in cold or in hot water. In cold washing they are dipped two or three times, and then allowed to stand and sweat to soften the dirt. They are then rubbed with the hand and passed through clean water. In hot washing they are first dipped two or three times in cold water to soften the dirt, then washed in a large tub in warm water with soap, and lastly, douched in cold water.

The Merinoes are a highly cultivated variety. The true Merino is of a fair size, and derives its distinguishing characteristics from the head, horns, fleece, and general contour. The head is very handsome, and horned in both sexes. It is Roman; short and broad across the poll or crown, and covered with wool over the ears and nose. The ears are short, and the horns must be open and wide between, well turned and marked with fine transverse wrinkles. The nose is often pink, but is better dark. The neck is short but full, and gains much character from the heavy folds or wrinkles of skin which adorn both males and females. The shoulders are very broad over the tops, and some are apt to be high or pointed in the withers. The body is long, the ribs deep and well sprung; the hind quarters apt to sink. The legs are short, and the hocks are sometimes narrow or cat-hammed. There are also folds of skin gathered together over the tail, giving the puckered appearance known as the "rose."

The Hungarian Merino wool ranges from one inch to one and a half inch in length, is of marvellous fineness, and mostly of rich orange colour, from the grease. The colour occurs in deep bands, which shade into a light yellow, and the samples have a rich candied appearance, as though they had been immersed in sugary syrup.

The Saxon Merino sheep have been divided into Rambouillet,

Negretti, and Electoral, but crossing has produced many varieties. Saxon Merino rams clip from seven and a half to fifteen pounds, and ewes five and a half to six and three-quarter pounds of wool.*

The Electoral Merino is well and strongly built: the head is of middle length and pretty broad, the neck short and fleshy, the shoulder and rump wide, back straight, and the body round; the feet are firmly placed and well set. The animals have thickly set

RAMBOUILLET-NEGRETTI RAM, POMERANIA.

wool, and are remarkably well covered, especially upon the belly, feet, and head. The wool is usually soft, of middle length, with mild and not too rich grease.

The Zackel sheep are distributed over the mountainous regions

* In a framed case, with 24 compartments (No. 10), will be found a good collection of Saxony wools, and another in No. 25. Case 52 contains 18 samples of fine German wools, and No. 54 samples of fine French wools and rovings. Cases 53, 55, and 56 contain fine samples of Hungarian wools, washed and in the grease, with locks mounted to show the staple.

of Transylvania and Galicia. They represent the division of long-woolled sheep (*Ovis strepsiceros*), are of various colours, and not fixed in their character even with regard to horns; some rams having long horns, while others, from Galicia, are hornless. They yield from six to eight and a half pounds of coarse unwashed wool. It is reckoned that each season the lamb brings four shillings, the wool four shillings, and about fifteen pounds of cheese gives four shillings more, making in all twelve shillings profit.

AFRICAN FAT-TAILED SHEEP.

There is a peculiar broad-tailed sheep met with in Asia Minor and Africa, which deserves to be incidentally noticed. In some parts of these countries, for example, in Cappadocia or Caramania, they have innumerable troops of sheep, which differ from those of Europe, by having a tail which is a mass of fat, and has often to be supported by a species of carriage. This variety has been known from the time of Herodotus. From Constantinople to Smyrna, and in all the adjacent districts, these sheep are eaten, and the tail is reckoned a peculiar delicacy. The wool, however, is coarse. This sheep is less common in Egypt, but is found in many parts of South Africa, among the Dutch. In the East, where pork is abhorred, this tail-fat is used, instead of lard and butter.

The sheep of the Kokan territory in Central Asia are of this large description, with heavy tails, which weigh, according to Mr. Michel, from 20 to 40 lbs.! They require little or no tending, are satisfied with the scantiest food, support thirst for a long time, and follow without fatigue the trails of their wandering proprietors. Their flesh is the favourite food of rich and poor, their milk and the cheese made from it takes the place of bread with a large portion of the population, and their skins form the winter garments of the people.

Our colonists in Australia and South Africa have proved the truth of the remark of an old writer on agriculture, Fitzherbert, that "sheepe is the most profytablest cattell a man can have."

Previous to the year 1833 the only sheep, with few exceptions, found in South Africa, were the broad-tailed, coarse-woolled animal. But in that year a few Merino sheep were brought over from Australia by a trading vessel, and were found much superior to the fat tails, and admirably suited to the country. Now the pure Merino, Cheviot, Escurial, and other esteemed breeds, are commonly to be seen in all parts of the South African colonies. The proportion of improved sheep to coarse-woolled in the Cape Colony in 1875 was as follows: 10,064,289 fine-woolled to 944,050 common or coarse-woolled.

THE WOOLLEN MANUFACTURE.—Wool is the second great textile industry of this country. There are more than a quarter of a million operatives engaged upon it in 2,500 factories; while the total number of persons directly dependent upon the trade may be set down (including the factory hands) at fully 1,000,000, there

FLOCK OF SHEEP IN AUSTRALIA, UNDER A LARGE EUCALYPTUS.

being a larger number of dependent workers in auxiliary trades than in connection with any other British manufacture.

	lbs.
We imported of Foreign Wool in 1874 . . .	338,800,481
Of Alpaca and Mohair	12,200,087
Woollen rags to be used as Wool. . .	57,361,920
Woollen Yarn for Weaving	13,114,130
Our home supply of Wool is estimated at . .	160,000,000
The Skin Wool from imported sheep at . . .	2,400,000
Making a total of . .	583,876,618

For the wool imported we paid more than £24,000,000 sterling. As we re-exported only 55,300,000 lbs. of wool and woollen yarn, nearly 530,000,000 lbs. of wool were worked up. It is estimated that we use at home three-fourths of the whole manufactures; therefore, as our exports of woollen and worsted manufactures and yarn in 1874 were of the value of over £28,000,000, this would bring up the aggregate value of the woollen manufacture to over £100,000,000, but even if we assume it at £75,000,000 it shows the importance of this great industry. The consumption of woollen goods for personal wear cannot certainly be less than £1 per head of the population, exclusive of the many other uses for woollen fabrics.

VALUE OF THE EXPORTS OF WOOL AND WOOLLENS FROM THE UNITED KINGDOM IN QUINQUENNIAL PERIODS.

Year.	British Wool.	Yarn.	Woollen Manufactures.
	£	£	£
1840	330,246	452,957	5,327,853
1845	556,339	1,066,925	7,693,117
1850	623,915	1,451,642	8,588,690
1855	986,523	2,026,095	7,718,374
1860	877,082	3,578,088	12,158,710
1865	901,660	5,110,474	20,141,415
1870	575,583	4,994,249	21,664,953
1875*	928,264	5,099,307	19,406,336

* The value of our exports of woollen manufactures and yarn in 1875 was nearly £4,000,000 less than in 1874, owing to the dulness of trade generally.

Sheep's Wool.—There are two broadly distinguished classes of wools that enter into commerce, fleece wools and skin wools. The first are obtained by the annual shearing of the sheep, the last are those cut or pulled from slaughtered animals, which, having been subjected to lime, are characterised by their harshness, weakness, and incapacity of taking a good dye, especially if the animal has perished from any malignant disease.

A knowledge of the qualities of wool is very essential to those engaged in the wool trade, which now forms an enormous business, nearly all concentrated in London.*

The fineness of wool is in direct relation to the thickness of the skin; the less thick the skin the finer is the wool which it secretes. But it is extremely difficult to obtain this product at the same time in large quantity, and also equally difficult in the Merino race of large size and great weight, to find the skin as fine as in smaller sheep. In increasing by an abundance of food the size of a given race, we increase at the same time the dimensions of the skin, as much as we thicken the surface.

The several qualities of wool are estimated with considerable accuracy by the cloth manufacturer, the wool sorter, and wool broker, who, by multiplied trials, have become experienced

* The collection of raw wools in the Museum is very large and extensive in its range. Attention may be specially drawn to the following among numerous other samples:

11. A series, in 330 bottles, of European, Asiatic, Australasian, African, and American wools, obtained from time to time at the various International Exhibitions, and from other sources.

On the east wall—

12 to **24.** Thirteen frames of sheep's and goat's wool from North America.

30. Samples of fine sheep's and Angora goat's wool from the Cape Colony.

31. Frame with fine wools from South Australia, accompanied with photographs of the sheep.

Case **49.** Samples of wool from Iceland and Madeira.

Case **51.** Eighteen samples of fine wool.

Case **57.** Illustrations of Smith's chemical process for cleansing burry, or foul wools.

in discovering, by the touch, minute differences quite imperceptible to common observers, and not even appreciable under the microscope.

The finest quality of wool is found upon the spine, from the neck to within six inches of the tail, including one-third of the breadth of the back or saddle. The second quality covers the flanks, and extends from the thighs to the shoulders. The third covers the neck and rump; and the fourth lies upon the lower part of the neck and the breast down to the feet, also upon a

DIFFERENT QUALITIES OF WOOL FOUND ON THE SHEEP.

part of the shoulders and the thighs to the bottom of the hind quarters. This is the Spanish and German classification, and is shown by the figures on the sheep above. Case 48 in the Collection may also be consulted.

A broad distinction is made in wools, which are divided by the trade into long or combing wools, applicable for stuffs and worsted goods, and short or clothing wools for cloth manufacture. The former are, however, again sub-divided into wools of from four to seven inches in length, used for hosiery and some other purposes, and those above that length used principally for coarse worsted goods. Specimens of the spinning processes of worsted yarns are well worth notice, showing that 89,000 yards may be spun to the pound weight, which is a great effort for worsted yarn.

In Germany the sheep are sorted and classed according to the fineness, length, and thick growth of the wool, and the following points are considered important :—

Strength of Fibre.—This is indicated by the amount of grease in the wool—plenty of grease indicating strength. This exists in three forms : soft or liquid, which again may be a rich yellow or white ; middle fat, yellow and white ; and, lastly, brown stiff fat, yellow and white. Of these, the oily or liquid grease is considered best in Germany, while in Hungary the middle fat is more suitable to the climate.

Fineness.—The wool should be equally fine over the whole body, but a coarser quality may be expected on the top of the shoulders and rump, and a weaker quality on the belly. It is with regard to fineness that the usual continental classification of Prima, Elector, Super, &c., is made. The finest samples of wool are usually not more than two finger-breadths in length ; but the *length* of wool varies from one to about four finger-breadths. The Prussian and French Rambouillet are longer woolled than the Hungarian Merino.*

Curl.—This is important, and refers to the minute bends or crimps which are seen in each hair ; a long, straight, plain wave in the fibre, as in B. (illustration page 28), is not liked, neither is an abrupt close wave which folds back upon itself, as in C. The best and most approved curl, that which gives spring and elasticity, as well as preserves the strength of the wool, may be described as a minute and regular serration, and is shown by the line marked A.

* Report upon the Vienna International Exhibition, 1873.

Thickness.—This quality refers to the thickness of the wool upon the skin, and is closely connected with the presence of those wrinkles so characteristic of the Merino sheep. Large folds of skin appear about the neck, and just above the tail in the true Merino, and especially in the rams. Young lambs, however, show the same peculiarity; and while, the wool is short, similar but smaller wrinkles over the entire body. The whole skin is completely furrowed with these wrinkles, and, consequently, the wool-bearing surface is rendered very large. It is considered in Austria a point of excellence when these wrinkles are numerous, but in Saxony a different taste prevails. The wool on the summit of

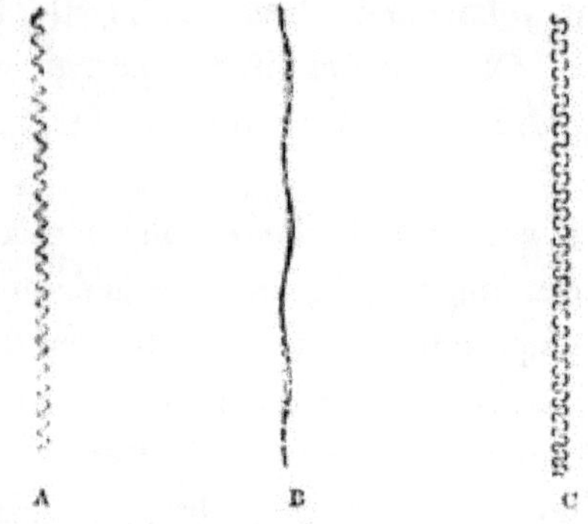

CRIMPS OF THE FIBRE OF MERINO WOOL.

folds seen about the neck is a little coarse, but as the area is small, this is not allowed to be a fault, only a character. A pure Merino sheep will carry from 40,000 to 48,000 wool fibres on a single square inch of skin.

The *closure of the stubble*, or outer surface of the fleece, is very important, for, if the fleece is loose and open, dust and dirt find their way into the wool. The closure is effected by the abundance of the fat, which rises to the surface of the fleece, and then mats the ends of the wool-fibres together, forming a compact protection to the fine wool beneath. This is further added to by dust, which adheres to the grease, and makes that firm black limit to the fleece always observable in the Merino. The hand passes over the stubble as over a sort of scale armour, and when pressed the springiness of the wool is at once perceived.

Opening a fleece for purposes of inspection is to be done with

"BLUMEN," OR FLOWER, IN MERINO FLEECE.

knowledge, and indicates at once whether the operator is at home with his subject. Grasping the points of the fibres with both hands, the inspector parts the wool and discloses the beautiful white or rich yellow, or orange-coloured wool below, and then closes up the fleece again without allowing any of the stubble ends to find their way down into the clean wool. The accompanying sketches illustrate the effect, which is very striking, when a fleece is opened, and also what is called the "*Blumen*" or flower (see p. 29), when the wool is made to open like a cup and exhibit its rich yellow and white colouring right down to the skin, reminding the observer of a fine lily.

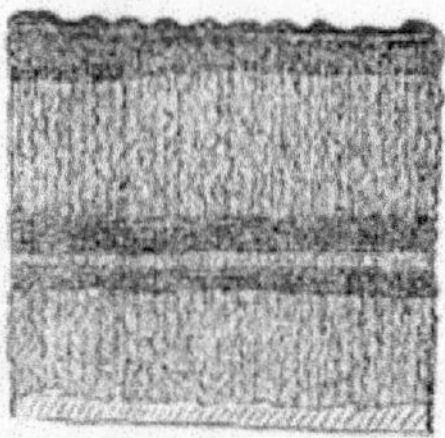

SECTION OF STUBBLE.

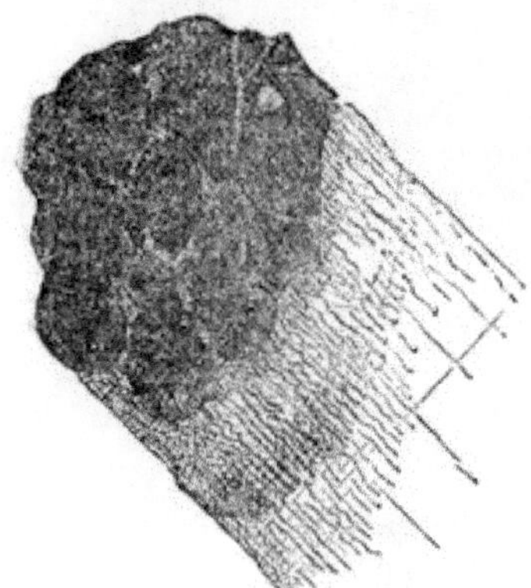

WELL CLOSED STUBBLE.

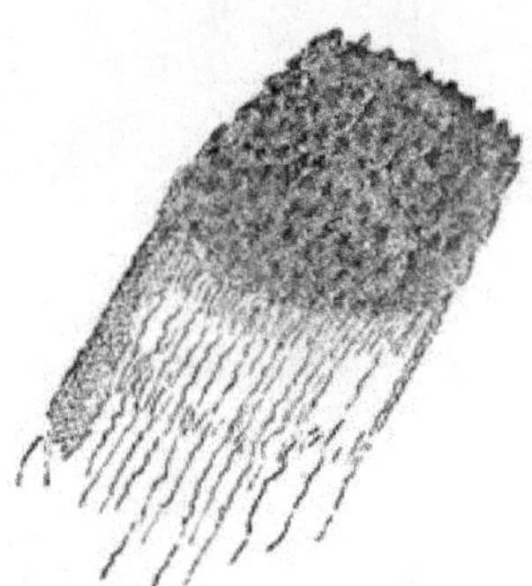

BADLY CLOSED STUBBLE.

A well closed stubble has the ends of the wool gathered into large masses, and has as few openings or crevices into the fleeces as possible. It is called a bad stubble when only a few fibres are caked together, giving the appearance of small dots instead of large bold blotches.

Growth.—The wool must be equally grown over the carcase. It must be equal in length on back and sides and belly.

The head must be woolled down to the nose and over the entire ears, and the legs must be clothed with wool down to the hoofs. Fine wrinkles on the horns are looked for in selecting rams.

On British sheep-runs there is some little difference in the classification, as the following remarks will show. The qualities considered most valuable in regard to the fleece are fineness, fulness, freeness, soundness, length, and softness.

1. *Fineness* of the fibre of the wool can be judged of by practice, when a lock of it is laid on the cuff of a coat of a dark colour. A deficiency in this quality will show itself by an abrupt falling off in fineness, either in the neck or breech of the animal, or in both. The difference in fineness between these parts and the rest of the fleece should be so gradual as to be almost imperceptible. No hair must be anywhere visible on the animal, especially under the forelegs.

2. *Fulness* means the closeness with which the staples or locks of wool grow together on the skin. Upon opening the wool of a sheep possessing this quality in perfection, only a thin line of skin, as fine as a pencil stroke, will appear round each staple, but if deficient, a space almost bare. Some of the German sheep, as shown in the illustration of the Negretti Merino ram (p. 32), have great rolls or puckers of skin under their necks or on other parts, which give them a singular appearance, but the extent of wool-bearing surface is thereby increased.

3. *Freeness* means that the separate fibres of each staple are distinct, and by no means entangled together, or what is called "smushy," like cotton wool. A deficiency in this quality shows itself most plainly along the ridge of the back. In a well-bred sheep the wool on being opened should fall apart under the hands as clear and broken as the leaves of a book.

4. *Soundness* or strength of fibre. Along the ridge of the back there is a sort of division between the wool of each side. Tenderness, that is, deficiency in soundness, invariably shows itself there.

Take out a staple from this part, and give it a strong steady pull, holding one end in each hand. If this proves sound, depend upon it that the whole fleece is so too. This is an indispensable quality in a combing wool, as there should be an absence of breaches or withered portions.

5. *Length* of fibre must be carefully regulated by the nature of the pasture and climate; for any the least excess, will cause a

NEGRETTI MERINO RAM.

proportionate deficiency in soundness, by which the wool will be depreciated for clothing, and rendered useless for combing. To judge of the length of the staple in a fleece, the best part to examine is the division along the ridge of the back, as it is there usually somewhat shorter than in other parts.

6. *Softness* sufficiently explains itself. A want of this quality is most conspicuous between the points of the shoulders and up the neck. Harsh wiry wool is more brittle, and suffers greater injury than soft wool in the various operations.

The way then to judge wool on a sheep's back, if it is really a fine wool, is, first to examine the shoulders as the part where the finest and best wool is usually found. This we take as the standard, and compare it with the wool from the rib, the thigh, the rump, and the shoulder parts, and the nearer the wool from the various portions of the animal approaches the standard, the better.

First, we scrutinise the fineness, and if the result is satisfactory, we pronounce the fleece in respect to fineness very "even." Next, we inquire into the length of the staple, and if we find that the wool on the ribs and back approximates reasonably in length to that of our standard, we again declare the sheep, as regards length of staple, true and even. We next desire to satisfy ourselves of the density of the fleece, and we do this by closing the hand upon a portion of the rump and of the loin wool—the fleece at these points being usually the thinnest and most faulty—and if this again gives satisfaction, we signify the fact by designating the wool "even" as respects density.

Now to summarise these separate examinations: If you find the fleece of nearly equal fineness from the shoulder to the thigh, of nearly equal length on shoulder, rib, thigh, and back, and of like density on shoulder and across the loins, you may conclude that you have a perfect sheep for producing valuable wool. A comparison of these two lists of desiderata, British and Continental, will enable a fair judgment to be formed of the quality of wool.

In the examination of wool the following points have also to be considered: the degree of imbrication of the surface as shown by the microscope; the quantity of fibre developed in a given space of fleece; the freedom of the fleece from burrs and other foreign matters; the skill and care employed in the scouring and other processes of preparation.

"Kempy" wool is objectionable, and the term means the presence of short white hairs at the roots of the staple, which never take the dye, and disfigure all goods into which they are introduced. The hairy East Indian wool that is usually grown near the tropics, has a tendency to be burry and scurfy, with a slight mixture of grey hairs.

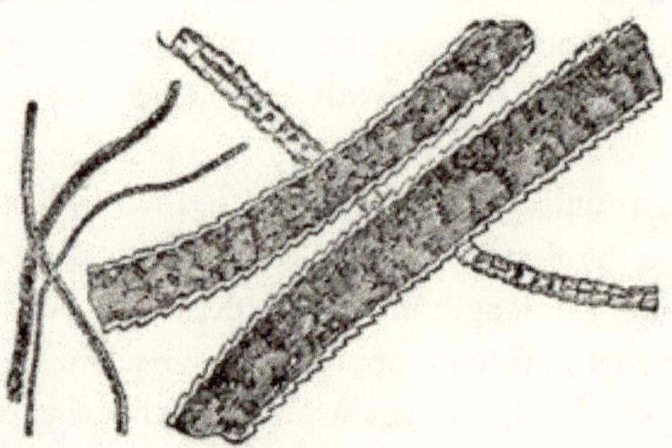

EAST INDIAN WOOL.

There is, perhaps, no defect which renders wool, and otherwise good wool too, so absolutely useless for manufacturing, and especially for combing purposes, as tenderness and breechiness. This fault, which causes the staple to be tender, arises from the destructive effects of drought, cold, or other climatic causes, which check the growth of the grasses and deprive the sheep of their necessary regular supply of food. But nothing is so sure to cause a break in wool, and in many sheep a perfect stripping or shedding of the entire fleece, as want of water.

It is not only important that wools should be free from the defects above described, but it is desirable that the whole of the various parts of the fleece should have, as nearly as possible, a uniformity of character, that is, as regards fineness, length of staple, density, and softness.

The illustration on the opposite page represents the different characters of the wools chiefly utilised: No. 1 being Cape sheep's wool, No. 2 Spanish Merino, No. 3 Southdown, No. 4 Camel's hair, No. 5 Mohair, or fine goat's wool, No. 6 Alpaca, No. 7 Llama, No. 8 that of the Yak.

In sorting wool for market, the fleeces should not be broken, but merely divested of the breech and stained locks, and assorted or arranged so that each bale or package may contain fleeces of the same character as to colour, length of staple, fineness of wool, and general quality.

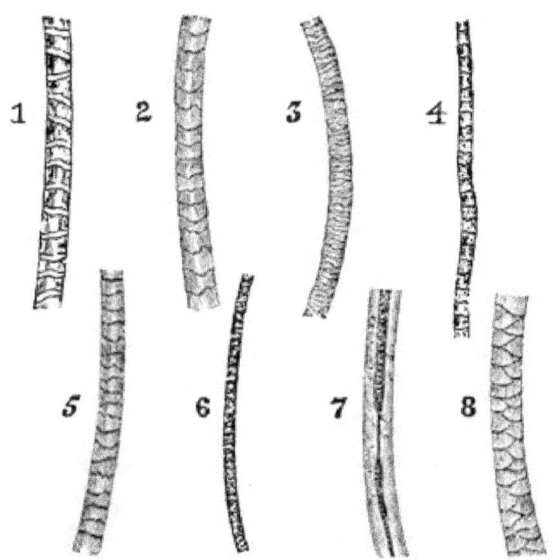

VARIETIES OF WOOL AS SEEN UNDER THE MICROSCOPE.

The wool from different parts of the same fleece is of various qualities, yet so sensitive by use do the fingers of the experienced stapler become, that it is with surprising rapidity he separates the masses before him into ten, twelve, or more kinds, taking due cognisance of the strength, cleanliness, regularity, colour, and softness of each. The fleece is sorted into combing and clothing wools, and broken fleece, or "pieces and locks."

In dividing and stapling, the fleece is spread by the fellmonger and laid on the sorting board, sheared side down. To the unaccustomed eye it looks a fleece of wool, all of one colour and quality, but to the sorter the different qualities are widely distinct. He breaks the skirts for one sort, the flank for another, the middle of the back and fore shoulders for another; and when the sheep has been crossed too broadly between coarse and fine, takes out the coarse piece which remains on the upper part of the neck, running up between the ears.

The classification terms and names for these several kinds of wool vary in different localities.

The following are the Yorkshire terms for "clothing" sorts, the first four being the choicest.

1. Finest picklocks.
2. Prime and pick.
3. Choice.
4. Super.
5. Headwool.
6. Downrights.
7. Seconds, from the throat and breast.
8. Abb, or an inferior sort.
9. Long livery, coarse belly wool.
10. Livery.
11. Grey.
12. Short coarse, from the breech.

The classification of the woolstapler is, to a certain extent, founded upon the difference of quality of wool arising from the parts of the body whence it is obtained; for example, that on the sides of the neck and shoulders, the ribs and back, is, as has been already stated, the finest part of the fleece; and next to this comes that which covers the thighs. But difference of race has also to do with the classification, for the wool on the breast of one sheep may be equal to that on the back of another; hence it is often not so much the object to separate the wools of the different parts of the body, as to put all the wool which may be adapted for one particular purpose by itself. A good fleece would generally come under the first four divisions, although portions might even belong to the sixth or seventh.

The deep "combing" sorts of wool are thus classed:

Long coarse.
Saycast.
Lusty.
Long Neate.
Long Drawing.
Fine Drawing.
Country Long Drawing.
Country Fine Drawing.

The following is the subdivision of a Southdown "tegg" fleece (that is the first shearing) weighing 6 lb. 12 oz., separated into eight qualities.

1. Super wool, 1 oz. ; used for flannels, blankets, hats, tweeds, and coarse cloths.
2. Livery wool, 1 oz. ; for low cloths, as prison, army, navy, and workhouse cloths.
3. Grey wool, 2½ oz. ; used for the same purposes, and hat making.
4. Prime white wool, 5¾ oz. ; used for cloth of all kinds, the best blankets, flannels, tweeds, shawls, Cobourgs, &c.
5. Choice wool, 2 oz. ; used for flannels, cloths, blankets, tweeds, and shawls.
6. Pick tegg wool, 1 lb. 7 oz. ; used for tweeds, shawls, and blankets.
7. Super tegg, 6¼ oz. ; for fringe and hosiery, yarns, and coach lace.
8. Long wool, 3 lb. 8 oz. ; used in yarns, fringes, shawls, blankets, &c.

The skin wools, or those from the slaughter-houses, have other curious terms, as lusty, kindly, ordinary, broad head, pick-lock head.

The wool of the lamb is generally softer than that of the sheep from the same flock, and as it has the felting quality in a high degree, is much used in the hat manufacture for the foundation or conical cap. The wool of lambs that have died a natural death possesses less of this felting property, and is employed for flannels and lambs'-wool hosiery. Young sheep's wool, and all long-grown staple wools, are bought by those who comb them for bombazines, camlets, etc. The short-stapled and weak-grown old sheep's wool can only be used by manufacturers of broad-cloths and fancy goods.

The great thing for promoting the growth of good sound wool is regular and generous feeding of the sheep, which insures a good supply of yolk, without which the wool would not possess elasticity, strength, softness, etc.

Sheep Shearing.—In our pastoral colonies, and in countries where large flocks of sheep are kept, the task of shearing is an important one. For instance, in the single colony of New South Wales, it was stated that in 1875, 25,000,000 sheep had to be sheared, yielding approximatively 125,000,000 lbs. of wool, and valuing this at 1*s*. per lb., it would amount to six and a quarter millions of pounds sterling. The cost of shearing this vast lot of sheep at 20*s*. per hundred—about the average price—would be £250,000, and that of transmitting the wool to the seaport for exportation may be set down at about the same figure. Without going into more minute details, if we estimate the value of the wool clip of New South Wales for 1876 at six and a quarter millions sterling, and assign 25 per cent. of that amount as expenses incurred by the wool-grower from the time the sheep enters the wool shed to be shorn (this is the estimated cost in the working of a wool station) until the net proceeds are in the wool-grower's bank, there will be disbursed £1,562,500. That sum would go in shearing, carriage to seaport and to London, commission, brokerage, &c. Every year Australian wool is increasing in quantity and rising in quality, so that at the close of 1880, New South Wales ought to have at least 30,000,000 of sheep, which, with horned cattle and horses, would approximately represent in money value upwards of £50,000,000 sterling.

Wool Washing.—Wool is sent to market in two forms, either in the grease or scoured; some manufacturers prefer to buy the former kind and wash or scour it themselves.

The great object to be obtained in washing wool is not only to make it white, but to render it bright. After washing the sheep with soft soap and warm water, avoiding all alkalies, which destroy the fibre (make it harsh and dry, "work unkindly," as the manufacturers term it), the fleece when squeezed by the hand should puff out again, not feeling sticky, and should glisten in the sun with a peculiar brilliancy; if too little yolk, or natural grease, is left in the wool, it will be wanting in softness; if too much, it will become sticky, and after a time turn yellow. The desirableness of this brilliancy in the wool is, that French manufacturers of merinoes, de laines, and other light textile fabrics, will give extreme prices for it, for only this bright wool will take delicate dyes.

Machinery is now applied for washing fleece and skin wool.

SHEEP-SHEARING OPERATIONS IN AUSTRALIA.

The ordinary process of washing wool is pretty familiar to most persons interested in that branch of industry, but the introduction of simple and effective mechanical appliances has thrown the former primitive method into the background.

The centrifugal pump of Messrs. Easton and Anderson (p. 41) is now much used in the Australian colonies for spout washing.

The apparatus is extremely simple, and consists of an iron tank mounted on a framing about eight feet from the ground, kept supplied with clean water by a centrifugal pump, driven by a portable engine. The tank is fitted with mouth-pieces, which are under control, and supply a torrent of water as required.

Another tank of wood, or any convenient material, about three feet deep, built in the ground, partly filled with water, is the "soaking" tank. The water in this tank is kept at about 69 deg. of temperature, and this is regulated by a steam pipe from the portable engine. The plan of proceeding is as follows:

The sheep are placed in the soaking tank and rubbed over. The warm water softens all the dirt in the wool. The sheep are then handled under one of the "torrents" from the upper tank, and the loose dirt is effectually washed away.

So speedy is the process that three sheep may be washed in two minutes, and so effectually that the value of the wool in the London market is very much increased. The waste in the soaking tank is about one gallon for each sheep.

It should be observed that the cost of the apparatus is the only outlay, no expensive material being used in the washing.

By a new chemical process, the sheep-skins in the tanyards are now stripped of their wool in an astonishingly rapid and effective manner, and the whole process, from the introduction of the raw skin into the place, to the dispatch of the well pressed bale, is interesting, from the clean, regular system in force.

A most ingenious pulling machine has been invented for clearing the wool off the skins. It is composed of a large revolving drum, driven by a belt, the motive power being steam. The

SHEEP-WASHING IN AUSTRALIA.

drum takes two skins at a time, and presses them under a nicely adjusted knife, which does its work most efficiently at the rate of 300 per day, or as much as six men could do in the same time. Moreover, the pelt is in no way damaged, and the appliance requires only the attention of two boys, who can stop the movement in an instant.

Wool and hair can be felted, that is, made into a dense and compact cloth, without the intervention of the processes of spinning or weaving. So great is this tendency that in a flock bed the carded wool, of which it is made, is constantly felting itself into lumps, and from time to time the bed requires to be taken to pieces, that the wool may be carded afresh. This felting property of wool and certain kinds of hairs, is caused by the peculiarity in the structure already mentioned; the filaments are notched or jagged at the edges, the teeth or imbrications invariably pointing upward, that is from the root to the point, so that the fibres, when subjected to gentle friction, move in one direction only, and consequently mat together and form the kind of cloth called felt. This felting property of wool is greatly assisted by the peculiar crimp in the fibre, which it retains with great pertinacity, so that if drawn out straight it immediately contracts again on being released; thus the forward motion of the fibre under friction is partly counteracted or converted into a circular or zigzag movement, which is precisely that which most completely effects the matting together of the various fibres.

Wool in the yolk, that is with the natural grease adhering to it, cannot be felted, the roughness of the fibre being in that case smoothed over by the oil; were it otherwise the wool would felt on the sheep's back and be comparatively useless.

For manufacturing, it is necessary to remove all the animal oil in order that the wool may take colour in dyeing; whilst a Cotswold fleece will lose in scouring but 18 or 20 per cent., some Merinoes will shrink 70 per cent. One great reason why the English and German woollen manufacturers beat the Americans in the brightness of colour and beauty of finish of their goods is, because they use more washed wool and less of the greasy wool.

The fleece of the Merino is so compact on the animal that the grease or yolk does not escape, but is condensed in the wool and produces a gum which is impervious to moisture, and while it protects the carcase from rain, its compactness causes all the more perspiration, which is produced at the expense of the food consumed.

The quantity of suint or yolk varies in different wools according to their fineness, in about the following ratio per cent. :—

Fine Saxon Electoral wool	80
French fine wools of Brie . . .	60 to 75
Merino wool	66
Common wools, rarely less than	20

CHAPTER II.

Treats of the several uses of Sheep skins, describes the woollen, worsted, shoddy, carpet, felt, and other British manufactures of wool, and gives the latest official statistics; proceeds on to the Goat tribe, from which we obtain skins, the materials for the principal glove manufacture, mohair, Cashmere shawls, and other useful products; and then follows on with details of the Alpaca, Llama, and Camel, and their various uses to Man, for their hair, flesh, and as beasts of burden; with full statistical information relating to the animals, and their manufactured products.

SHEEP-SKINS.—There is an extensive use of sheep and lamb-skins for different purposes. About 17,000,000 are obtained annually from home-slaughtered animals, and 10,000,000 more imported from abroad. They are usually split into two portions, known respectively as "skivers" and "fleshes," the former being the grain side; the unsplit skins are termed "roans."*

Sheep-skins form a large item in the commerce of the Cape Colony. The shipments there now reach about 1,500,000 skins annually. It is chiefly, however, the indigenous or half-bred sheep that are killed, the Merinoes being more valuable to keep for their wool.

The manufacturers of boots and shoes consume large quantities of sheep-skins for linings, toppings, etc. Roans are also finished to

* In frames hung on the raised divisional partitions will be found a series of samples, furnished by Messrs. Bevington and Sons, of some of the uses of sheep skins tanned, for coloured roans, furniture morocco, hard and cross-grained morocco, hand-grained skivers, hatters' skivers, coloured calf, grained sheep, hard and cross-grained sheep, glazed sheep, cross-grained sheep, cross-grained and glazed sheep. The whole process of preparing and tanning sheep skins is also shown in a series of large photographs of the various departments in the Bermondsey tannery of that firm.

imitate goat-skin morocco to a considerable extent. Bookbinders use very much imitation morocco made from skivers, and so closely does this resemble the real article that, when on the book, it takes a good judge to detect the difference; "fleshes" are employed in the manufacture of blank books, and for other purposes, where strength rather than superior finish is desirable. Sheepskins are also used by trunk and bag makers, saddlers, pocket-book manufacturers, jewel-case makers, hatters, organ and other musical instrument makers, upholsterers, glovers, suspender manufacturers, druggists and perfumers, etc. In England quantities of fleshes are manufactured and sold as "chamois skins." The glove manufacturers are also large consumers of fleshes, having them dressed in oil and finished to imitate buck or deerskins.

Sheep-skin mats are prepared by stretching the fresh skin, well furnished with a coat of wool, with the wool side down, by means of tacks at the edges, and then rubbing the skin over with a powdered mixture of equal parts of common salt and alum—repeating the operation twice afterwards on the two following days. It should remain exposed to the air, but not to the sun, till well dried. When skins are "tawed" with the wool on, as for mats and rugs, they are doubled with their wool side inward, so as to expose only the flesh side to the alum mixture.

"Slink" lamb-skins are called in Persia Karpak. They are greatly prized, and fetch as much as seven pounds for ten or a dozen skins. These skins are obtained by causing the ewes to bring forth prematurely, which is managed as follows: The ewes, when within a month of lambing, are driven out two miles or so on a cold night, and are brought back suddenly into a very warm stable. The violent change of temperature causes the ewes to bring forth, and the skins of the lambs thus born are those which are so highly prized. A hat made from these skins will sell for nearly £10, and in such a hat only the prime parts of the choicest skins will be used.

In some northern countries the ewe is killed purposely to

obtain this fine skin wool from the unborn lamb, in order to be made into glove linings and for saddle covers, &c.

The Russian, Astracan, Hungarian, and Spanish lamb skins are remarkably fine.* The grey and black Russian lamb-skins are mostly used for cloak and coat linings, collars, caps, &c. The Astracan lamb has a rich, glossy, black skin, with short fur, having the appearance of watered silk. The Astracan lamb-skins used to be much in demand in China, as many as 1,750,000 having been shipped via Kiachta in some years; now scarcely 100,000 are sent yearly. The Hungarian lamb-skin is used in that country in immense numbers; of it is made the national coat. In summer the woolly part is worn outside, in winter inside. They are often highly decorated. The Spanish lambs furnish the well-known short jackets of the country.

WOOLLEN MANUFACTURE.—The Woollen Manufacture begins with the stapler, who buys the wool of the farmer or wool-broker, and ends with the merchant. It is divided into three principal processes, which are again subdivided.†

First there is what is called the Manufacturer;

Secondly, the Finisher; and

Thirdly, the Rag Grinder.

The first manufactures the raw material into cloth.

The second finishes it, and gives it its appearance as it is ordinarily worn.

The third takes the manufacture of the two former processes, when thrown aside by the wearer, cuts it into patches, which he forcibly tears asunder into woolly fibres, and then remodels this again into the raw material known as shoddy, to be once more used by the manufacturer.

And of so much consequence is this last process to the trader,

* Samples of these will be found in Case 97.

† The whole stages of progress of the woollen manufacture, the various kinds of woollen goods, and the shoddies, are all shown in the Animal Products Collection of the Bethnal Green Museum.

that there are machines in Yorkshire capable in full work of producing 50,000,000 lbs. of raw material per annum, or (upon the average of 6lbs. to the fleece) of adding to the annual stock of wool the fleeces of more than 8,000,000 sheep. Besides tearing up the old worn garments at home, we also import annually 57,000,000 lbs. of foreign woollen rags to be thus torn up and reworked.

The object of the shoddy manufacture is to supply cheap and economical clothing to all, especially to the working and poorer classes. This is effected by utilising materials which previously were considered almost valueless. Shoddy cloth is generally used for winter wear. The raw material employed in its production is shoddy or mungo, in combination with wool, noils, or waste. Rarely is any cloth made of shoddy alone. The chief seat of this important and flourishing business is Batley. It is also extensively carried on at Dewsbury, and prevails in the district generally. Shoddy is "rag wool," or wool produced from rags. The rags are ground or torn up into "wool" or "flock" by a cylinder set with sharp iron teeth, and revolving at a rapid rate. Shoddy is the produce of what are termed "soft rags," such as cast-off woollen stockings, flannels, shawls, carpets, stuffs, &c.; mungo, that of cloth of a finer texture, such as dress coats, tailors' clippings, &c. These are ground up in the same way as the old stockings and flannels, but they require a finer set cylinder and a little different treatment. The rags are obtained from everywhere, being collected by bagmen, marine store dealers, &c., and thus heaps are accumulated. The leading provincial towns furnish considerable quantities, but London is the great depôt. Rags are received, too, from foreign parts, especially from the Continent. A very large quantity was formerly imported from Germany; but they are now mostly ground up there, reaching this country in the form of shoddy and mungo. Shoddy varies largely in value; some varieties are sold at under 1*d.* per lb., others at 9*d.* or 10*d.* per lb., according to colour, quantity, and staple. Mungo varies still more, namely, from 1*d.* or 1½*d.* per lb. to 20*d.* and upwards.

The average price of shoddy is probably about 4*d.* per lb.; that of mungo about 6*d.* The sheep's wool required in the manufacture differs greatly in price; ranging from shearlings at 4*d.* and coarse Scotch at 6*d.* per lb., to fine Australian at 2*s.* 6*d.* and upwards. Noils and waste are obtained principally from Bradford, where spinning and power-loom waste are produced in considerable quantities, and though a refuse in the Bradford trade, serves a useful purpose in the shoddy manufactures. The price of shoddy cloths vary in price according to weight and width, from 1*s.* to 12*s.* per yard. There is a collection of shoddy and mungo ranged in 102 bottles near the wools, and they are also shown with the cloth made from them in the Waste Products Collection.

Another use for waste wool is to grind up very fine, and, when dyed of various brilliant colours, it is sifted or powdered over fresh-varnished paper-hangings, to which it adheres, forming the elegant velvet or flock papers imitating figured tapestries and stuffs. A collection of bottles of these "flocks" is placed on shelves near the shoddies on the south side of the gallery.

Three great divisions of the trade are commonly recognised—the manufacture of Woollen cloth, of Worsted or stuff fabrics, and of Hosiery. The chief districts in which they are carried on are as follows:—Woollen cloth in Yorkshire, Gloucestershire, Somerset, and Wilts; stuffs or worsted at Bradford, Halifax, York, Leeds, Worcester, and Norfolk; hosiery in Leicestershire; woollen yarn in Lancashire, Derby, and Yorkshire. Besides which carpets are made at Kidderminster, Wilton, and Axminster, and tweeds, plaids, and woollen shawls in Scotland.*

* In Cases **50**, **58**, **59**, and **60** will be found an instructive series of samples, showing the several stages of preparation from the raw material to the yarn dyed and undyed, and the finished cloth. In Cases **62** to **67** are samples of broad cloth, livery, and other woollen fabrics, and the various fancy woollen goods made in different foreign countries. Case **75** is devoted to flannels, machine blankets, and roller cloths, which are exported to the value of about £1,500,000. Case **47** shows the wool sorted into classes for this purpose. Case **77** contains ruggings, coverlets, and wrappers, which we ship to the

WORSTED MANUFACTURE.—The term "Worsted Stuffs" is applied to those manufactures into the composition of which wool enters that has undergone the processes of combing, and includes those fabrics in which cotton and silk are combined with combed wool. The name "Worsted" is derived from a village in Norfolk where these goods were first produced. These fabrics are carefully to be distinguished from woollen cloths, the chief characteristic being, that they undergo the well-known process of "felting" or "fulling." Though for some centuries the worsted manufacture had its chief seat in Norfolk, Suffolk, and Essex, it has now obtained a remarkable concentration in the West Riding of Yorkshire, the town of Bradford being the principal centre.

The latest official returns give 630 factories engaged on the worsted trade in the kingdom. These employ 111,000 operatives, and have 2,160,000 spindles and 65,000 power looms.

Of the total yearly value of the worsted manufactures produced it is impossible to form anything more than an approximate estimate. The export of worsted goods, mixed and unmixed, has in some late years reached a value of nearly £21,000,000, and the exports may be roughly taken at from one-half to three-fourths of the whole production. When we compare the value of the exports of worsted stuffs in 1855, the sum being then under £4,000,000, the enormous advance made in the worsted industry of the country is evident.

CARPET MANUFACTURE.—Another branch of the woollen trade

value of £250,000. Case **68**, mixed tweeds and ribbed cloths, and Case **76**, Scotch Tartans. Case **80**, woollen mantillas, sashes, scarfs, &c., made in Spain. Case **91**, fine woollen and mixed fabrics for ladies' dresses. Case **81** contains Barège and other woollen shawls, made for the Great Exhibition of 1851, which are characterised by their elegance of design, variety of patterns, and excellence of manufacture, for which the town of Paisley is noted. Case **82** contains white and grey woollen yarn, and specimens of under-clothing and hosiery. Case **79**, illustrations of the process of combing wool and yarn of different degrees of fineness. Case **81**, a lady's boa made of fleecy wool from Thanet. No. **69** is a large embroidery picture representing the Lord's Supper, showing the use of Berlin wool.

may be studied, that of carpets, of which there are numerous illustrations of the stages of progress and the varieties of fabrics in Cases **71**, **72**, and **73**. The more general use of carpets at home and abroad has given a great impetus to the carpet trade; the demand both for home use and export having greatly increased. Twenty years ago we only exported carpets of the value of less than £500,000, now the shipments of carpets reach a value of £1,500,000, and those used at home must be fully double that value or more. The long combed wool is principally used for Brussels and the finer kinds of carpeting; while carded wool is employed for Scotch, Kidderminster, or common carpets. Brussels carpeting has for its basis a warp and woof of strong linen thread. Tapestry and other common carpets have a backing of yarn made of cotton waste.

There are several other varieties of carpets made; Kidderminster is the chief seat of the manufacture, but Halifax, Durham, with Kilmarnock, Dundee, and other towns in Scotland now compete.

Druggets or felt and jute carpets have become an important item in the trade.

The manufacture of carpets is now principally effected by means of machinery: hand-loom Jacquard weaving having been almost entirely superseded by steam-power weaving.

Leaving now the consideration of the Sheep, we pass on to notice some other wool-bearing animals and their commercial products.

THE GOAT TRIBE.

Angora Goats.—An important wool-bearing animal of commerce is the Asiatic goat, which, unlike the common goat, has a long and silky white fleece, that is in great demand both locally and for export, and passes here under the name of "Mohair."

Twenty years ago there was a fine flock of 200 Angora goats in Spain, kept in the mountains of the Escurial, belonging to the State, and there are now many hundreds owned by private individuals there.

When some bales of this goat's wool were first shipped to England from Turkey in 1820, the article was so little appreciated that it brought only tenpence per pound, but the demand has been annually increasing, and it now fetches 3s. 6d. per pound. Our yearly imports of goat's wool average 7,000,000 pounds, of the value of about £1,000,000.*

Its chief value depends on the length and fineness of the staple, its bright and silky appearance, and its softness. It is extensively spun into yarns in Yorkshire, and the finer sorts are used in great quantities at Lyons in the manufacture of lace, and in the German States for dress goods and shawls.

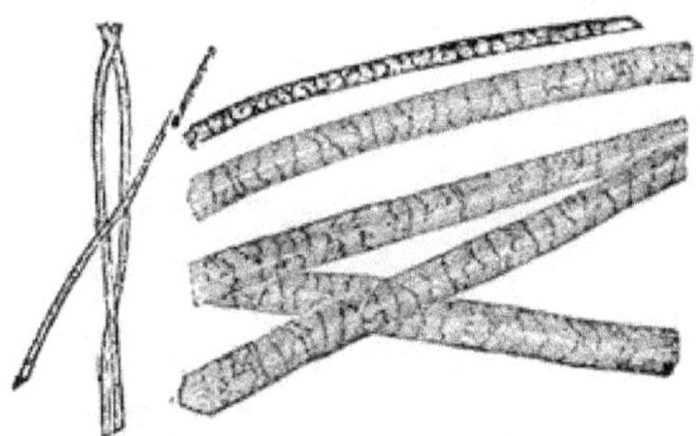

MOHAIR, OR GOAT'S WOOL.

The Cashmere, Persian, Angora, and Circassian goats are one and the same animal, changed in some respects by altitude, though but little by latitude. They abound in the inaccessible regions of those countries, and are the flesh, milk, cheese and butter-yielding and wool-furnishing animal of the whole country. They are finely developed for the table, much disposed to fatten, with very white and beautiful long fine wool or curly hair, the fleece weighing, in different districts, from two to three and a half pounds. The larger Kurdish goat with black wool has a more extensive range than the white goat, which is better known by the name of the town which is the centre of its range—Angora.

* Cases **89** and **90** contain samples of Cashmere and Angora wool, and other kinds of goats' hair; and Cases **88**, **91**, **92**, and **93**, fabrics made of it.

The Angora goat, although originally confined to a limited district, has now been transported to and reared successfully in many other countries. The fleece is locally called "tiftek." After the goats have completed their second year, they are clipped annually in April or May, and yield progressively until they attain full growth, from one to three pounds. The process is perfectly simple, the fleece being of pretty uniform quality, and, unlike the Thibet or Cashmere goats, which have a downy covering on the pelt, with long coarse hairs at the top, the separation of which is both tedious and expensive, the Angora goat's wool is packed and shipped as it comes from the animal.

A serious drawback to the development of the trade in goats' hair is the determined dishonesty of the native graziers and dealers, who persist in drenching the fleeces with water, and mixing in all sorts of rubbish in order to increase the weight, and so realize an extra profit, although by so doing they destroy the lustre of the hair, which is its principal recommendation to spinners, as it enables them to use it as a substitute for silk. The English agents in Angora have made repeated efforts to have the practice forbidden, and have even induced the Central Government to interfere to prohibit it, but the active opposition or passive resistance of the local authorities have hitherto frustrated their wishes.

The Kirghiz of Kokan keep large flocks of goats, which in character are not unlike those of Thibet, with reddish grey hair of great length, under which is a beautiful white hair of the finest description, from which the inhabitants of Urutupa manufacture shawls and scarfs as fine and as highly prized as those of Cashmere.

Fine goats' hair is produced in the Kirghiz steppes, and sells in Khiva and Bokhara at 28*s.* the cwt. Thirty years ago about 80 tons of goats' down, and about half that quantity of goats' hair, used to be annually shipped from Russia. Now only about 12 tons of it are imported annually across the Orenburg frontier. Russia exports the great mass to Europe; but a portion of it is manufactured in Eastern Russia into socks, gloves, girdles, and shawls of great variety and surprising brightness.

The Angora goat is now successfully reared in South Africa, South Australia, Victoria, and other of the Australian colonies. There are nearly 1,000,000 in the Cape Colony, where they

ANGORA GOATS.

were only introduced about ten or twelve years ago. As early as 1848 they were imported into South Carolina, and have since spread over many of the States of the Union, the original stock being surpassed in beauty and amount of hair or wool produced, which is partly due to the extreme care taken in breeding, and partly to the colder winters on the North American Continent, which tends to increase the woolly covering of quadrupeds.

Since 1870 a large number of pure bloods have been imported into California, and they are now quite common throughout the States and territories of the Pacific, where their number is estimated at about two millions, and it is doubling every year. They thrive much better in the Pacific States than in the Northern and Southern States. These animals are very prolific, and if well kept have kids when only one year old. It is also said that the flesh of the Angora goat is far superior to that of the common goat, and even better than mutton, veal, or venison, according to the testimony of some who have lived on it in California.

The southern slopes of the Himalaya Mountains afford the most congenial locality for the famous shawl-wool goat. The northern face of these mountains is as remarkable for its dryness as the southern is for its moisture; the cold is excessive, and the animals which are pastured there are covered with shaggy hair, or with long wool and a fine down. Few are aware of the tedious protracted labour attending the manufacture of a fine Cashmere shawl. When the hairs have been separated, the residue is carefully washed in rice-water and hand-spun by women, who do not earn more than equal to about half-a-crown per month. There necessarily is, in the manufacture of these shawls, a great division of labour. One artizan designs the pattern, another determines the quantity and quality of the thread required, while a third arranges the warp and woof. Three weavers are usually employed on a shawl, of only an ordinary pattern, for three months; but a pair of rich ones not unfrequently occupies a shop, or a family, for a year and a half. They are dyed in the yarn, and carefully washed after the weaving is completed. The Cashmerian dyers of eminence profess to produce sixty-four tints, some by extracting colours from European woollens. The embroidered borders of the finest shawls are invariably made separate, and afterwards skilfully sewed on to the main piece. The immense labour required to produce a first-rate Cashmere shawl, will account for the fact, that a shawl will sometimes cost £600 or £700 before it passes the rocky portals of the valley of Cashmere. These shawls always form part of the presents made to distinguished persons who visit the courts of Eastern princes.

The Pashum, or shawl wool, properly so called, is a downy substance found next the skin and below the thick hair of the Thibetan goat. It is of three colours, white, drab, and dark lavender. The best kind is produced in the semi-Chinese province of Turfan Kechar, and exported *via* Yarkand to Cashmere. All the finest shawls are made of this wool; but as the Maharajah of Cashmere keeps a strict monopoly of the article, the Punjab shawl weavers cannot procure it, and have to be content with an inferior kind of pashum, produced at Chathan, and exported *via* Leh to Umritsur, Nurpur, Loodianah, Jelapur, and other shawl-producing towns of the Punjab. The price of white pashum in Cashmere is, for uncleaned, 3*s.* to 4*s.* per lb.; cleaned, 6*s.* to 7*s.*; of the Tusha, or dark lavender wool, 2*s.* to 3*s.* per lb., uncleaned; and 5*s.* to 7*s.* cleaned.

The shawls of Kerman, in Persia, are not much inferior to those of Cashmere. They are woven by hand, similarly to the carpets. The material called "koork," of which the shawls are made, is the under wool of the white goat, numerous flocks of which animal are in the neighbourhood. These flocks migrate annually, according to the season. Major Smith tells us that he made inquiries at Kerman why the "koork" producing goats were only to be found in that neighbourhood, and was informed that in that district the rapid descent from the high plateau of Persia to the plains near the sea, afforded the means of keeping the flocks throughout the year in an almost even temperature and in abundant pastures, with a much shorter distance between the summer and winter quarters than in other parts of Persia, and that such an even climate, without long distances to traverse in the course of migration, was necessary to the delicate constitution of the animal, or rather to the softness of its wool. The whole of the "koork" is not made use of in the looms of Kerman, a large quantity being annually exported to Umritsur, in Upper India, where it is manufactured into false Cashmere shawls.

The Common Goat and its Commercial Products.—Being the natural inhabitant of mountainous regions, and injurious to trees and shrubs, it is in wild, rocky countries that the goat is mostly reared. In Europe, the goat is chiefly found in Spain, which has about 4,500,000; Portugal and Austria 1,000,000 each; Germany nearly 1,700,000; France 1,800,000; and Russia 1,400,000, besides some in Asia. The United Kingdom has about 1,000,000; Sweden and Norway 500,000, and smaller numbers are found in Italy and Switzerland. Asia maintains a large number, for in the Peninsula of India alone there are about 6,000,000.

Statistics of Goats in various countries according to the latest official returns.

Country	Year	Number
Russia	1871,	1,330,000
Sweden	1873,	121,840
Norway	1875,	357,102
German Empire	1873,	1,709,521
Saxony	1867,	93,003
Holland	1873,	146,169
Belgium	1866,	197,138
France	1872,	1,791,725
Portugal	1852,	1,014,742
Spain	1865,	4,531,228
Italy	1874,	1,688,478
Austria Proper	1871,	979,104
Hungary	1870,	459,810
Switzerland	1866,	375,482
Moldavia and Wallachia	1873,	305,316
Greece	1867,	2,289,1[illegible]3
Turkey	1874,	1,500,000
United Kingdom (estimate)		1,000,000
Morocco, 1875 (estimate)		12,000,000
Egypt	1871,	23,97
Algeria	1861,	3,500,000
Cape Colony	1875,	3,095,41
British India (estimate)		6,000,000
Ceylon	1872,	88,197
Réunion	1866,	11,322
Martinique	1865,	4254
Guadaloupe	1866,	9336
United States	1875,	2,500000
Argentine Republic	1874,	1,300000
Uruguay	1871,	60000

The wool-bearing goats of Eastern Asia and their products have been already alluded to, but it may be mentioned here that there are about 1000 pure Angoras in Cordova, South America, introduced from the Cape Colony in 1864, and 2000 to 3000 of various crosses, besides a million or more of native goats, which can be bought there for about 3*s.* each.

Between Kangra and Ladakh, goats and sheep are much used

for light burdens, especially in the rice trade between Kooloo and Ladakh. Flocks of these hardy little animals are daily to be met skipping over the rough and rocky roads, and up and down their precipitous sides, with loads of 24 to 32 lb. on their backs, and they travel 12 and even 15 miles a day without any apparent labour or fatigue.

Passing now to Africa, in the countries bordering on the Mediterranean, Morocco, Tunis, Algeria, &c., there are many millions of goats, and at the southern extremity of the continent, the goats number about 2,500,000. Although goats are numerous in Morocco, especially in the southern part, the skins exported annually, only average about one million and a quarter, a large number being used up locally. The tanners of Europe have, however, learned to excel the Moors themselves in the art of preparing this kind of leather. In Morocco it is made only in four colours; bright yellow, which is largely used for men's slippers, white and red for women's slippers, and brownish-red, which is employed for other purposes. From 250,000 to 300,000 pairs of these Moorish slippers are annually exported.

In Northern Africa, three races of goats are chiefly met with, the Bedouin, or true Arab goat, the Maltese, which has long been introduced, and the Spanish goat, which is met with in some parts of Algeria.

The Bedouin goat is of medium height, well formed and hardy. The hair is generally black and long, and is used for various purposes; mixed with sheep's wool or camels' hair to make cords or fabrics for tents, and alone to stuff cushions. The chief value of the goat to the Arab is, in common with the sheep, to furnish milk, as drinkable water is often difficult to obtain, especially in the south. The goat is also used as a beast of burden in some districts to transport water in goat skins.

Smaller than the Arab goat, the Maltese goat is a much better milch animal, frequently yielding five or six quarts of excellent milk per day. The colour of its hair, like that of most domesticated animals, varies considerably. Unlike the Arab goats, which are mixed with the sheep, the Maltese goats are kept in separate flocks on the hill slopes in the neighbourhood of towns, and are valued at about £3 or £4 each.

The Spanish goat is larger and stronger built than either of the preceding, and is a good milker, but not equal to the Maltese variety. Its hair is black or brown.

The goat is also spread over many parts of America, especially in the Northern and Southern regions.

The mountain goat (*Aplocerus montanus*) is found throughout all the mountain ranges of North-West America, to within a short distance of the Polar Sea, if indeed it does not reach it. It is a larger animal than the domestic goat (*Capra hircus*), which it only resembles in name, and in having a beard. It is covered with long and rather brittle white hair, beneath which a coat of very fine white curly wool lies close to the skin. The flesh, though rank, is fat and tender, and is much relished by the mountain Indians, who also make robes, clothing, and leather from the skin.

In mountainous countries goats render considerable service to mankind, the flesh of the old ones being salted as winter provisions; they are the cow of the poor, and their milk is used in many places for the making of cheese, as in Portugal, where four or five million pounds are made. The flesh of the kid is highly palatable, being equal in flavour to the most delicate lamb. But it is for the kid skins—for glove making—that the animals are most valued. The importance of the glove trade may be estimated by the French manufacture, which is said to exceed 30,000,000 pairs, of the value of between £3,000,000 and £4,000,000 sterling, giving employment to nearly 90,000 persons. Although many of these gloves are made of lamb and other skins, the best are always of kid.

Glove Manufacture.—Independent of the elastic quality of the skin, a good glove is distinguished, first, by its being neatly sewed, secondly, by the thumb seam not extending too far into the palm, and lastly, by the colour of the exterior not having soiled the inside. Most of the lower priced English gloves offered as kid, are in reality made of lamb skin. Beaver, though the quality is various, forms the commonest description of leather gloves. The Woodstock is a very superior beaver, to which much attention is paid both to the shape and sewing. Doeskin is a more thick, durable, and soft leather. Buckskin is the closest grained, and consequently the stoutest leather of which gloves are made. Its elasticity, though trifling, is sufficient. It also bears cleaning better than any other kind. It may be had in white, drab, or buff. Sheep skin is generally white, and most usually made by contract for the army. Tan is of three qualities, and is a very serviceable and cheap glove for gardening, riding, or driving. The home production of gloves was valued at upwards of £1,000,000 sterling a quarter of a century ago; but there are no reliable details to guide us as to the value of the manufacture at present, although it has been stated at £2,000,000. Austria has of late years made great strides in glove-making, for the exports of gloves are now valued at £4,500,000. Our foreign imports of gloves are large. In 1850 we imported 3,250,000 pairs of gloves. In 1860 the duty was removed, and the quantity of foreign gloves received has since largely increased. In 1874 we imported 13½ million pairs, valued at 1½ million sterling. The manufacture is principally carried on in the towns of Worcester, London, Yeovil, Stoke, and Milborne Port. Woodstock and Witney are known for the so-called "beaver," made of buck and doe leather for military and hunting purposes. In Hexham and Nantwich a trade still lingers for tanned gloves. What are called "dog-skin" gloves are an English speciality made of Cape sheep skin. Limerick is very famous for its gloves, which are so soft and light, that a pair are frequently packed in a walnut shell as a present. The French glove manufacture is an important and celebrated one. Nearly 2,000,000 dozen pairs are annually made, valued at about £3,000,000, of which three-fourths are exported. The principal towns where the trade is carried on are Paris, Grenoble, Chaumont, Milhau, Nancy, and Blois.

It was only in the reign of Louis XIV. that leather gloves began to be made in France.

Gloves play a conspicuous part in many national customs which originated in the days of chivalry, such as throwing down the glove as a challenge of combat, hanging out the glove at licensed fairs, &c. The origin of the custom which prevails at a maiden assizes, when a pair of gloves is presented to the presiding judge, is not generally known. In former times the judges of the land were prohibited from wearing gloves on the bench, and hence the custom, when there is no judicial business to discharge, of presenting a pair of gloves.

As the delivery of gloves was once a part of the ceremony used in giving possession, so the depriving of them was a mark of divesting a person of office, and degradation. Challenging by the glove was continued down to the reign of Queen Elizabeth in this country, and is still in use in some parts of the world. One ceremony yet remains with us, in which the challenge is given by a glove, at the coronation of the sovereign, when the champion, completely armed and mounted, enters Westminster Hall, and professes his readiness to meet in single combat, any one who disputes the title of the successor to the crown, and thereupon throws down his glove or gauntlet, as a token of defiance.

As the leather for gloves requires to be very soft, it is prepared by a process called tawing. Two kinds of skin are employed for conversion into the better qualities of kid leather, one of these, the most expensive, being the skins of young goats, fed solely with milk, the other lamb skin. Each of these skins yields on an average two pairs of gloves. After being soaked in water to soften them, scraped on the "beam," buried for a fortnight, and soaked in a fermenting mixture of bran and water, the skins are steeped for a few minutes in a mixture of alum and salt dissolved in water, or in this solution with flour and yolk of egg; they are then washed and dried. The yolks of eggs are added only in the preparation of the finest kid leather. The yolk of egg acts by the oil it naturally contains in the state of emulsion, this oil giving to the kid leather that suppleness and softness which is so much esteemed in gloves. One hundred skins require from six to eight pounds of alum and as much salt, and one egg for each skin. The dyes are applied either by immersion or by brushing over the leather ; the latter, or English method of dyeing skins, is more ordinarily practised. Sheep skins are split into two sections, and the flesh side serves for military gloves. Lamb skins are too thin to be split, but are made into what are called beaver gloves. Very good white kid leather is obtained by tawing the epidermis from lamb or goat skins in a saturated solution of stearic acid in alcohol. The leather thus obtained is very soft, has a whiter colour than ordinary glacé leather, and a beautiful gloss.

The use of gloves has become so general an article of dress among all classes of society, that besides the very large import of foreign made gloves, our home trade supports a vast number of men, women, and children, who are employed in their manufacture ; the cutting out, sewing, binding, setting on the buttons, lining, and trimming in large manufactories, affording as many different branches of occupation.

In Case **139** will be found collections of gloves, and lamb and kid skins prepared, showing the mode of cutting out gloves, also dressed goat and sheep skins, cat skins, and gloves made from rat skins.

Goat skins are tanned and made into various kinds of leather, for ornamental purposes.

The importance of goat skins is shown even by the British commerce in them. We imported in 1874 about 7 millions of goat and kid skins, of the value of £850,000 ; of those imported $5\frac{1}{2}$ millions were prepared in some way, either tanned, tawed, or dyed. The enormous progress that has been made in the trade is evidenced by comparing the number imported in 1855, which was only 1,200,000 skins. We draw our supplies mainly from two quarters:—5¼ millions of goat skins reach us from India, chiefly from the Madras presidency; and 1 million from South Africa, where there are now about 3,000,000 goats of all kinds.

Twenty years ago we scarcely received a quarter of a million from South Africa. Goat skins are received brined or salted. The largest and best of the latter kind average 52 to 56 pounds weight per dozen; seconds, 42 to 45lb. the dozen; and small, 28 to 34lb. the dozen. East Indian tanned skins ranging from 9 to 12lb. weight are worth 13*s.* to 15*s.* per dozen. Best grain goat skins 40*s.* to 80*s.*, and seconds 18*s.* to 48*s.* per dozen.

The hair of the goat makes good linseys, and that of the Welsh he-goat is in request for forensic wigs. Ropes are sometimes made from goat's hair, and are said to last longer, when used in water, than those of hemp. Coloured goats' hair from the tanneries sells at about 2½*d.* per lb., and white fetches 6*d.* per lb. It is used for making yarn, coarse blankets, mops, &c. Candles are made with goats' tallow, which, for whiteness and quality, are said to be superior to those of wax. The horns of the goat afford excellent handles for knives and forks.

Skin bottles, which were used before the discovery of glass and cask-making, are still much employed as a means of transport in Southern Europe, parts of Asia, and Africa. In Spain a wine skin made of goats' hide or hog skin, pitched or rosined, is found more convenient for carrying on the back of a mule, and is cheaper than a cask. The Arabs always use leather bottles made of goat skins for their water, milk, wine, &c. Some of these are shown at the Bethnal Green Museum, hung under the stairs leading to the Picture Gallery.*

When the goat is killed, the head and lower portions of the legs are cut off; then without cutting the skin, except a little at the neck, it is drawn off the body. It is then cleaned and

* There are water skins, butter bottles from Egypt, and many large leather tobacco pouches shown.

tarred; if it should be used without tarring, an unpleasant taste would be imparted to the liquid. When thoroughly prepared, the skin at the leg ends is tied up, and the neck, which is the mouth of the bottle, is also furnished with a string for tying.

Leather bottles containing fifty or sixty gallons are also made of the skins of oxen. One of these, filled with liquid, is a load for a camel. The camels, in returning, can carry back other articles of merchandise, while the merchants, taking their empty skin bottles and filling them with air, tie a great number of them together in the form of a raft. Then laying a few boards on the top of these, they have a vessel with which they can navigate the stream, and placing on it some of their attendants, with such goods as they wish to send home in this way, they dismiss them. Thus we see the leather bottles are sometimes put to a double use.

Here is another account of the mode in which the leathern bottle was, and is still, made in France, Spain, and the East, according as it is sewn or not sewn. To make a bottle which is *not sewn*, a buck is taken and bled to death, then inflated with the traditional bellows to detach the skin from the flesh. The head is then cut off above the neck, and the fore-legs at the knee joints, after which the carcase is hung up by the hind legs, and the whole body is got piecemeal out of the opening at the neck. This done, the hind legs are cut off, and the skin is turned inside out; the latter is then cleaned in various ways, but especially with pounded salt, which has the property of preserving it. It is then placed under a stone, and left there for several days, fifteen at least. When it is calculated that this summary preparation has succeeded, the skin is again turned and shorn more or less closely according to the quality. Lastly, with catgut or simple twine, the openings of the legs and neck are sewn up, unless the latter opening is preserved for the purpose of closing it eventually with a bung surrounded with linen, and in which a hole is made of the diameter of the cork or other stopper intended for it. As for the opening of the anus this must perforce be sewn up.

The *sewn* leathern bottle is generally made of cow hide; skins without folds are chosen, and these are soaked in lime water which has already been used. As soon as they are sufficiently softened, they are put into a bath of fresh whitewash; then they are peeled, washed in running water, then dried, at first in the sun, and afterwards in the shade on dry ground. The skins are then put again into water, and, lastly, cut into shape and sewn by the aid of a cobbler's awl, the flesh side inwards. It is curious to see with what rapidity in certain countries the workmen will cut out and sew up these leathern bottles.

There are several ways of preserving them for a very long time, and of preventing them from wearing into holes; the most prevalent method, especially in the case of those made of buck skin, is to treat them with honey-water, in which has been dissolved some rye-flour carefully sifted.

There is a large trade done in leathern bottles in the Levant, especially in towns where caravans and armies are provisioned. In the Turkish army, a certain number of soldiers, in regular companies, have to carry these bottles filled with water, to serve out to their officers and comrades in the field. The leathern bottle held its ground in France for a long time against clay and wooden vessels, and even barrels; as late as the time of Philippe of Valois leathern bottles were seen on the king's table. By reason of the difficulty of gauging their contents, however, they could never be used as measures; but when a bottle of any liquid was mentioned, the precaution was always taken to specify the quantity it should contain. The Arabs, however, make these bottles of a stated capacity, and they contain as near as possible the quantity of liquid which it is declared they can and ought to hold. They are often made double; this, however, does not prevent the scorching winds of the deserts from drying them up completely in a few hours.

THE LLAMA TRIBE.

Of the animals of the Llama tribe of South America every part is turned to use. Their wool is made into cords, fabrics, and sacks. Their skin is tanned, their bones are used to make weaver's utensils, &c., and their dung is employed for fuel.

LLAMA.

There are four distinct species of this family common to South America, the Llama (*Auchenia Lama*), the Alpaca (*Auchenia Paca*), the Guanaco (*Auchenia Guanaco*), and the Vicugna (*Auchenia Vicugna*).

The llama is the largest and least valued of the domesticated animals, and, like the camel, is used as a beast of burden in Peru, but it is more useful than the camel, inasmuch as its flesh serves more generally for food, when young it is savoury and nutritious, and its wool for clothing and other useful purposes. The number of these animals employed as beasts of burden, in conveying produce, &c., from the interior to the coast, and between town and town, the mines, &c., has been estimated at nearly 2,000,000. Their load is from 60 to 150 lbs., and they travel about nine or ten miles a day; they are frequently thus driven in flocks of from 500 to 1000. Very little llama wool is exported, the demand being great for local consumption, the quantity used up for sacking, cordage, carpets, and fabrics being estimated at 5,000,000 or 6,000,000 lbs. The llamas used as beasts of burden are never shorn, their wool serving the purposes of a pack.

The most useful of these animals for its wool is the Alpaca. Its fleece is superior to that of the sheep in length and softness, averaging 7 to 9 inches, and sometimes it is procured of an extraordinary length. The fleeces, when annually shorn, range from 7 to 12 lbs. Contrary to experience in other descriptions of wool, the fibre of the alpaca acquires strength without coarseness, besides each filament appears straight, well formed, and free from crispness, and the quality is more uniform throughout the fleece. There is also a transparency, a glittering brightness which is enhanced on its passing through the dye vat. It is distinguished by softness and elasticity, essential properties in the manufacture of fine goods, being exempt from spiral, curly, and shaggy defects; and it spins easily when treated properly according to the present improved method, and yields an even, strong, and true thread. Notwithstanding the remarkable quality and beauty of the alpaca wool, it was long before its value was appreciated in Europe. Now the imports of this wool into the United Kingdom are from 4,000,000 to 4,500,000 lbs. annually, of the aggregate value of £550,000, the ruling price being, in 1876, about

2*s.* 4*d.* per lb. There is a good collection of alpaca wool and fabrics at the Bethnal Green Museum.*

To Sir Titus Salt, of Bradford, must, undoubtedly, be awarded the high praise of finally overcoming the difficulties of preparing

THE ALPACA OF PERU.

and spinning alpaca wool, so as to produce an even and true thread, and by combining it with cotton warps, which had then

* Cases **83** and **84** contain 16 samples of alpaca and vicuna wool of different colours from Peru; **86**, alpaca wool from South Australia and Victoria; **87**, alpaca wool and yarn; and **85**, three llama fleeces—grey, brown, and black—from animals raised in this country, and presented by Baroness Burdett Coutts. Case **88** has a progressive series, illustrating the stages of the alpaca manufacture presented by Sir Titus Salt and Sons: and **91**, **92**, and **93**, are cases of fine woollen mixed fabrics, alpaca, mohair, &c., for ladies' dresses.

(1836) been introduced into the trade of Bradford, improving the manufacture, so as to make it one of the staple industries of the kingdom. By an admirable adaptation of machinery the material is now worked up with the ease of ordinary wool.

Distinguished at first, when the Spaniards discovered Peru, by the name of "Peruvian sheep," on account of its woolly fleece, a better acquaintance with the Alpaca has proved that it is a widely different animal. Its size is somewhat above that of a large goat. It is reared for its wool, which is of much higher value than that of the sheep, and for its flesh, but little inferior to mutton, and of which it yields perhaps three or four times the weight that a sheep does.

The Peruvians long guarded the Alpaca with scrupulous jealousy, and endeavoured to restrict it to their own country by the most stringent prohibitions and penalties against its shipment. A few, however, are now scattered over Europe. In 1858 Mr. Charles Ledger succeeded in taking a flock of Alpacas through the interior of South America, and shipping them at a Chilian port for Australia, landing 276 at Sydney. Although the Colonial Government bought these, and paid great attention to them, they have not progressed so rapidly and so well as was at first anticipated. Like the Merino sheep in Australia, they may, however, form the nucleus of future prosperity, when we consider that in 1801 there were under 7,000 sheep in Australia, while the numbers in all the Australian settlements in 1875 were over 62,000,000.

So valuable are their services considered, that in the interior provinces of Peru there is scarcely a cottager who does not keep a dozen llamas and as many alpacas. The first serve to carry his corn, his potatoes, and his fruit to market, while the wool of the latter clothes his family, and the flesh of both, fresh or dried, affords a wholesome meal. Their skins also serve him for a bed.

The Guanaco is the largest of these animals, and ranges over a greater extent of country, being found on the immense tracts of table-land as far south as Terra del Fuego, and north to the slopes of the towering Chimborazo. Dr. Darwin mentions having met with herds of 500 in Patagonia. The wool, which is of a dark brown colour, is much shorter, coarser, and intermingled with hair. The Patagonians and Auracanian Indians of Chili work it up into blankets, ponchos, &c., while its skin is used as a quilt. We do not receive any large quantity of guanaco or vicuna wool. The flesh of this animal is the best of the class and is highly esteemed.

The Vicuna is the smallest but most graceful animal of any of the species, and its wool, of a pale reddish brown, more of the beaver caste, is finer and even more valued than that of the alpaca, but its yield is small, seldom exceeding a pound a year. The flesh salted, and dried under the name of "charqui," is eaten, but is not considered so good as that of the Guanaco or Alpaca. Opinions on this point seem however to differ, some considering the flesh of the vicuna equal to venison.

THE CAMEL AND ITS PRODUCTS.

There are two distinct species of Camel, the one-humped and the two-humped.

The common two-humped camel (*Camelus Bactrianus*) is the largest, and is spread over Central Asia, between the Sea of Aral to Siberia, Thibet, and China. It is bred far to the north by the Kirghiz and Cossacks. In Mongolia the traffic is transferred from the backs of this camel to those of reindeer from the Arctic regions. The two-humped camel has been known from ancient historic times. It is said to be still found in a wild state in Turkistan and Thibet, and is distributed among most of the tribes of Asia.

Throughout Turan the two-humped camel is in general use, from its being better able to support the cold than the dromedary, which is employed on the southern slopes of the Hindu-Kush. In Khiva, however, a large single-humped camel, called "Nar," is in great request. About 6000 camels are engaged in the caravan trade from Bokhara to Russia, conveying Asiatic goods. Laden camels travel the road between Yarkand and Kokand.

The camels employed in the caravans that cross the Siberian steppe to Russia carry loads varying from 640 to 720 lbs. In Persia there are a great many dromedaries, and there is an artillery corps of some 200, trained to carry a gun, gunner, and ammunition. Turkey also has a regiment of six squadrons of camels for military service.

The Arabian or single-humped camel (*Camelus Dromedarius*) is met with in Persia, Syria, Arabia, Egypt, Northern Africa, and Senegal.

The ordinary Arabian camel is not, properly speaking, the "dromedary," which name, of Greek origin, denotes a light and highly cultivated breed of racer proportions, celebrated for its fleetness, the *Mehari* or *Ashari* of Barbary, and known as the *Sindairi* in India, which is a very different race from the heavy baggage camel of the same countries. A fleet or dromedary race is said to exist in China.

Although the dromedary is a native of Arabia, it is in Africa, where the conquering Arabs have carried it, that it is the most multiplied, and its characters have been brought to the greatest perfection. The "mehari," or saddle dromedary, of the best breed and the fleetest pace, can exceptionally cover a distance of 120 miles in a day, and has done 200 in 24 hours without drinking or eating. According to the Arab saying, the mehari is to the djemel or pack dromedary what the noble is to the peasant.

Dr. Leared, in his recent work, "Morocco and the Moors," says that the burden camel, "Jimmel," is there the most important of the domesticated animals. By its means the products of distant provinces are interchanged, and commerce is carried on with places like Timbuctoo, in the heart of Africa. The strength and enduring qualities of the camel alone make such journeys possible. Day after day, from sunrise to sunset, this patient animal will plod through a desert at the rate of about two miles an hour, while carrying a load of four hundredweight or even more. To sustain all this patient toil, a small meal of grain, or even of straw, with water at intervals of days, will suffice. In the north, where the camel is larger, and food can be procured on a journey, the

THE MEHARI, OR RACING CAMEL.

animal will carry as much as six hundredweight. The ordinary price of one is about £10.

The fast camel, or "Mehari," bears the same relation to the burden camel as our thoroughbred horse does to a cart-horse. The form of the fast camel is more slender and elegant, and his special characteristic lies in his speed. The statements made in respect to this, as also of his endurance, seem almost fabulous. This breed of camels varies greatly in excellence, and one of first-rate quality is valued at a very large sum of money.

Camels have adaptations of structure to the peculiar mode of life, and to the climate and conditions of existence, in the localities in which they are appointed to dwell, the callosities enabling them to kneel or lie down on the burning sand without inconvenience, and the hump containing a supply of nutriment against a time of scarcity, analogous to the internal provision for storing up water which the camels have in common with the llamas. After a long and wearisome journey, the humps of the Bactrian camel more especially, are described by several observers to hang over like empty bags.

The provision for storing water is of two kinds,—the "honey-comb-bag" of other ruminants being converted into a receptacle for water, to contain any excess of fluid obtained by drinking, while a large cellular apparatus connected with the paunch retains the excess of moisture derived by secretion from over succulent food. The structure of a camel's foot—a broad convex cushion underneath—is especially adapted for treading on soft yielding sand, while the joints are so articulated that the foot is necessarily lifted high at every step, without wearisome effort to the animal. Again, the long pliable neck, the hard palate, and the powerful teeth, are obviously suited to seize and tear away and masticate the tough prickly vegetation of the desert; and the eyes, nostrils, and ears are beautifully guarded against the intrusion of particles of sand!

The one-humped camel of Turkestan is a very different animal from the Arab or Indian camel, and is even considered a distinct species by Professor Eversmann in his notice of the camels of Bokhara.

The Lohani merchants, who are called Povindahs or runners, and carry on the trade communication between India and Central Asia, have an immense number of camels. They number some 12,000 fighting men and 60,000 camels, and every year they lose a hundred or more men and at least two per cent. of their camels, besides some hundred loads of merchandise at the hands of Waziri and Sulairiman Khyl tribes. A direct trade was first opened up about thirty years ago between Troitsk, on the Orenburg frontier, and Chuguchak, across the Kirghiz steppe, by the enterprise of the Tartars, with a caravan consisting of 70 camels, each carrying a load of 560 lbs.

Camels' hair is produced in Bar and Thal, or waste tracts in Shakpur, Rohtak, Shang, and Gugaira, which are camel-feeding districts. The soft underdown, which is of a light brown colour, is used in the manufacture of cloth for "chogas" of a common kind. In former years a good deal of camels' hair used to be shipped from Russia, occasionally exceeding 27,000 pounds.

As for colour, the Indian camels vary from black to white, with every intermediate shade of cream, drab, and mouse colour; but the extreme colours are rare. In Arabia it is said that a lady of Nedj considers it a degradation to mount any other than a black camel; while an Ozanian beauty prefers one that is grey or white. In the continuation of Clapperton's journey, by Lander, we are told of the arrival of 500 camels laden with salt from the borders of the Great Desert, which "were preceded by a party of Tuarick merchants, whose appearance was grand and imposing. They all entered full trot, riding on handsome camels, some of them red and white, and others black and white."

A caravan in Africa may consist of a thousand camels, nay, sometimes four or five thousand collected together, and a single individual will be the master of four or five hundred. The Dey of Tunis singly owns thirty thousand. It is difficult to form an estimate of the number of camels, as precise data are wanting. Twenty years ago there were 60,000 in the Russian Empire, nearly all in Europe and the Caucasus; Spain has 3000 or 4000, chiefly in the Canaries; in Tunis there are probably 50,000, but in Africa and Asia there must be very large numbers. The camels in Algeria are estimated to number 180,000; in the single province of Oran there are more than 60,000. The camel arrives at maturity in about five years, and the duration of its life is from forty to fifty years, but it varies, in India seldom living longer than twenty-five years, and in Algeria and Egypt thirty.

The camel is an important adjunct to trade and commerce in tropical sandy regions as a beast of burden. Small camels will carry a load of from 600 to 800 pounds; while the largest and strongest will bear a burden of 1000 pounds or upwards at a rate of thirty to thirty-five miles a day. Camels used for speed alone will, however, travel more than double that distance.

The camel is of great service to the Chinese in the northern provinces, and to the Mongols, as a beast of burden; they eat its flesh and make ropes of its hair. Its fat, called "the oil of bunches," is used in rheumatic affections.

The natives of Africa esteem camel's flesh more than that of any other animal, but in other quarters it is not held in equal favour, being hard and unsavory, and little esteemed even by the Tartars. They use, however, the hump cut into slices, which, dissolved in tea, serves the purpose of butter. The camel was eaten both by the Greeks and Persians. Heliogabalus had camel's flesh and camel's fat served up at his banquets, and the flesh of the young dromedary is considered, by the Arabs, equal to veal.

Camels were introduced into Peru and Caraccas from the Canaries, at a cost of about £1400, in the middle of the 16th century, but did not thrive, and were superseded by mules and llamas. Camels have also been carried to Australia and Brazil at great expense, but have not succeeded well.

The two-humped camel may be classed among the wool-producing animals. The most expensive of the articles the Bokharians bring are the shawls, said to be fabricated of the soft downy hair of the dromedary's belly. They are made of strips about eight inches wide, sewed together so neatly, that in coloured goods it is impossible to discover the junction. The white shawls have a variegated border made of the fibrous cuticle of a plant of the nettle tribe. These are so highly valued, that the sum of 1200 roubles (probably paper roubles), £180, is sometimes paid for one.

Camel's hair, or more properly wool, for the hairs are in small proportion, is of great value in the countries where the animal is employed, but our imports are very small, only from 200,000 to 300,000 pounds annually; its principal use being for making fine brushes or pencils for water-colour painting.*

The quality of the hair of the dromedary varies on different parts of the body, the finest and best is on the shoulders and the hump. In the young animal the hair is fine and smooth, but becomes curly and crisp with age. The camel is shorn in the spring after the second year.

The following description shows the admirable processes of nature in adjusting the growth of the wool so as to suit the climate and the season of the year:—"The wool is so thickly disposed that the skin of the animal can with difficulty be discerned beneath it, even when the wool is turned back for that purpose. In the spring, as the temperature grows milder, the whole of this wool detaches itself from the skin, being pushed off in masses and flakes by the hair, which springs up beneath it, and which forms the summer clothing of the animal. It is at this season pulled or cut off, and after being cleaned, is either manufactured into woollens, of different textures, for home consumption, or exported in a raw state to Russia. This wool is called *koork*, or down, and appears to be little inferior in fineness to that procured from some

* In Case **86** will be found samples of camel's hair of various colours, and in Case **78** velvet cloth made of it, and fine French cashmere shawls made of it, mixed with the silky Mauchamp wool.

breeds of shawl goats, while it possesses a decided advantage over them all, in being both of a much longer fibre and far more easily freed from hair."

The quantity obtained from an ordinary camel weighs about 10 lbs., but its colour and abundance depend entirely upon the particular species of camel, and the climate which it inhabits.

The hair is prized according to colour, the black is the dearest, the red the second quality, and the grey is only worth half the value of the red. When spun it serves for wrappers for merchandise, and for making the tents, shawls, and carpets of the Arabs. In Persia more valuable manufactures are produced in cloths of different colours and fine stockings. The Tartar women of the plains make a kind of warm, soft, and light narrow cloth from the hair of the Bactrian camel, preserving the natural colours. The flesh and milk of the camel are a means of sustenance to man. Its skin is worked into leather, its manure is used for fuel, and it has been termed "the ship of the desert," carrying heavy burdens patiently.

The camel has long been acclimatised in Spain, and breeds there. There are many belonging to the government as well as to private individuals, and they are sold at from £15 to £20. They have not only been used there as beasts of burden and draught, but to plough and turn oil-mills. They are fed like other domestic animals, on straw, hay, oats, &c.

A good dromedary costs at Cairo about £8, and the price of the best kind of fast camels does not exceed £20.

CHAPTER III.

PRODUCTS OF THE BOVINE TRIBE.

Having noticed the Sheep, Goats, and other wool-producing animals, we are next led to consider the Bovines and their economic relations to man, as furnishing food, hides, tallow, horns, bones, and other products. In this chapter estimates are given of the cattle in the principal pastoral countries. The races of oxen, bison, and buffaloes, of different regions, are next described; their importance, both in the wild and domesticated state; the improvement made in the breeds in Britain and Europe; our meat supply from cattle; the comparative consumption of meat in different countries; modes of drying meat; the dairy products, milk, butter, and cheese, are briefly described, and statistics of production furnished.

CATTLE AS FOOD PRODUCERS.—Having dealt with the ruminating mammals which are of the highest utility as wool-bearing animals (chiefly those of the Ovine race), we come now to speak of the Bovines, which are primarily useful to man as food-producers, although furnishing at the same time many other commercial products of considerable value, in hides, tallow, horns and bones, &c. To civilised nations and many semi-civilised tribes, the possession of herds of cattle is of great importance. The pastoral wealth in some countries is indeed greater than can be profitably utilised locally, and the difficulty is, in such vast areas as Russia, parts of South America and Australia, to prepare the various flesh products in a form to be transported and profitably saleable in the densely populated States of Europe, where the demand for animal food and the raw materials for manufactures outstrip the supply.

The greater part of our domestic animals having been transported to America, have multiplied there prodigiously. The cow, the bull, the horse, the sheep, and the pig, are all species of the Old World, which were quite unknown in the New, and yet they may now be counted there by millions. So in the Australian colonies, where all these were only introduced less than a century ago, the pastoral wealth or live stock is still the leading characteristic of the various settlements.

Whilst the sheep will live and even fatten upon the poorest vegetation, the ox needs a richer and more nutritious diet to make flesh and fat for the butchers and the feast. In England, Holland, and many other parts of Europe, the ox feeds upon fat meadows where water is at hand.

Cattle are of value on more accounts than their flesh. The hide is an article of marketable price. Even the hoof and the hair have their uses, and hence are in demand in the great workshops of the world, where science and skill fashion into forms, available for some one or other of a thousand purposes, the material which, but for the thrift of modern industry, would be cast aside as waste.

The number of neat cattle belonging to the principal countries of Europe, according to the latest returns, is about 100,000,000. The States owning the largest number are, Russia, 23,000,000; France, 11,300,000; Austria and Hungary, 12,000,000; and the German empire nearly 16,000,000, of which Prussia has 8,600,000.

The United States have 27,220,000 cattle, and the Dominion of Canada, 2,700,000. The River Plate States have about 22,000,000 head, of which Uruguay has 7,000,000, and Venezuela, Costa Rica, Chili, Brazil, and other American States a great many cattle.

It is by thousands that the Tartars of Mongolia count their flocks and herds, and there are chiefs who own more than 15,000, distributed over various points of the immense steppes, and guarded and directed to fresh pastures on horseback. From Africa and the Eastern countries we have less specific details of numbers. But we know that both in Central and Southern Africa there are large herds of cattle.

CATTLE in various countries according to the latest official returns.

EUROPE.

Country	Year	Number
Great Britain .	. 1876,	5,848,214
Ireland .	. 1875,	4,111,990
Russia . .	. 1870,	22,770,000
Sweden .	. 1873,	2,183,394
Norway . .	. 1865,	950,000
Denmark .	. 1871,	1,238,898
German Empire	. 1873,	15,776,702
Saxony . .	. 1867,	625,260
Holland .	. 1873,	1,432,091
Belgium . .	. 1866,	1,242,445
France .	. 1872,	11,284,414
Portugal .	. 1870,	520,474
Spain .	. 1865,	2,904,598
Italy . .	. 1874,	3,489,125
Austria Proper	. 1871,	7,425,212
Hungary .	. 1871,	5,279,193
Switzerland .	. 1866,	993,241
Greece .	. 1867,	109,904
Moldavia and Wallachia .	. 1873,	1,886,990
Great Britain .	. 1876,	5,848,214
Ireland .	. 1875,	4,111,990

AMERICA.

Country	Year	Number
United States .	. 1875,	27,220,200
Canadian Dominion, and other British Colonies	1871,	2,724,760
Uruguay .	. 1872,	7,200,000
Argentine Confederation .	. 1875,	15,000,000
Falkland Islands	1873,	25,000
West Indies, — British and Foreign (estimate)	—	500,000

There are no reliable data for Brazil, Venezuela, and the other South and Central American States, but they possess a large number of cattle.

AFRICA.

Country	Year	Number
Egypt .	. 1871,	132,666
Algeria . .	. 1861,	1,053,086
Cape Colony	. 1875,	1,097,506
Dutch Republics and Kafirs (estimate)		1,000,000
Natal . .	. 1874,	501,154

ASIA.

Country	Year	Number
Mauritius .	. 1875,	29,545
Java . .	. 1873,	4,358,105
British India (estimate)		30,000,000
Réunion . .	. 1866,	6,000
Ceylon .	. 1873,	826,690

AUSTRALASIA.

Country	Year	Number
New South Wales	1875,	2,856,699
Victoria .	. 1875,	958,658
South Australia .	1875,	185,342
Western Australia	1875,	46,748
Queensland .	. 1874,	1,343,093
Tasmania .	. 1875,	110,450
New Zealand .	. 1874,	494,113
Hawaiian Isles	. 1866,	60,000

We have no reliable returns of the entire live stock of British India, but the number of horned cattle in the Ganges Valley must be enormous, judging simply from the export of hides from Calcutta, which occasionally exceeds six millions annually. When it is remembered that this quantity represents only the surplus stock that is left over from the Bengal Presidency after the wants of the entire native community have been supplied, we may safely assume that the total number of cattle equals, if it does not exceed, that of human beings in that part of India.

There are probably, according to a well-informed Indian paper, (the *Delhi Gazette*,) one hundred millions of horned beasts to be found between the Sutlej and Calcutta, a number which perhaps does not exist anywhere else on the globe, except in the pampas and prairies of North and South America. A more striking proof of the fertility of the Ganges Valley could not be given than the fact that, with a population per square mile greater than that of most European countries, it nevertheless supports a number of cattle only about a third less than is to be found in the whole of Europe.

There are now nearly twice as many cattle in the several British Colonies as there are in the mother country, for Australia and New Zealand possess 6,000,000; our African settlements and Indian Islands, 2,500,000; British North America and the islands of the west, 3,000,000; making a total of 11,500,000. In 1850 Australia only possessed 2,000,000 head of cattle, so that the number has trebled in a quarter of a century.

The Basutos, Kafirs, Fingoes, and other native tribes clustering about the eastern borderland of the Cape, own more than two million head of domestic animals, valued at £3,500,000, comprising more than half a million of horned cattle and 1,000,000 sheep. In the Dutch Free States and in Natal, there are also a large number.

There are four distinct kinds of horned cattle met with in Southern Africa. 1. A coarse-boned, long-legged breed, with enormous horns. This is best calculated for the yoke or "treking" work, on account of its activity, and is known as the "Africander" breed. 2. A more fleshy and thick-set animal, with smaller horns and softer hoofs. This was originally imported from Holland, and is in high esteem for milking; it is known as the "Fatherland" breed. 3. A diminutive active, and somewhat, humped animal, which is found chiefly among the natives, and which seems to have, with its masters, an inbred detestation of all kinds of artificial restraints. This ox is, in all probability, a cross between an Asiatic quadruped and a Spanish beast from the Portuguese South American provinces; the tendency to rise in the back, being derived from the eastern side of the parentage. This is known as the Zulu breed. 4. A long-legged animal, with remarkably poor quarters, and with horns even bigger than the Africanders. This belongs to the Basuto or "Macatees" breed. This animal is not often seen, and is of very low value indeed.

In the neighbourhood of Lake Tchad, and in the kingdom of Bornou, cattle are kept in great abundance. They perform all the laborious business at home of carriage and tillage, the camel only being used for war and extensive journeys. They are the bearers of all grain to and from the markets.

Major Denham in his Travels tells us that a small saddle of plaited rushes is laid on him, when sacks made of goat skins, and filled with corn, are lashed on his broad and able back. A leather thong is passed through the cartilage of his nose, and serves as a bridle, while on the top of the load is mounted the owner, his wife, or his slave.

The long inherited habit of the South African native is a delight in horned cattle. The habit has grown up from many motives. The natives are great milk drinkers. It is with cattle they buy their wives. And they have a gentlemanly liking for a fine animal, and especially for a swift racing ox. Then again a large, herd is a sign of wealth and respectability. It has not been the custom of the native to take a commercial view of horned cattle, unless in relation to wife-buying. But within the last few years a preference for sheep has shown itself, and on the sole ground of the profitableness of wool. The Kafir is actually beginning to barter away his beloved and cherished cattle for an animal which promises to be remunerative.

The ox-hide is of indispensable utility for many purposes, both in South America and in Africa. In the former it is the principal material for the packages called "serons," in which Paraguay tea, barks, medicinal roots, and various descriptions of produce, are transported.

In South Africa, raw hide is used by the Boers as a substitute for all kinds of cordage. It is made into drag ropes for the waggons, head-stalls for the oxen, bridles for the horses, cordage for thatching the hut, slips for bottoming the beds, chairs, and stools, pickling tubs for his beef, and "feldt schoon" for himself and family.

There are five great grazing regions in the world. First, the interior of Asia, which has furnished rich pasturage, summer and winter, since the time of Abel, who was a keeper of sheep. The second great pasture-ground is South Africa. That immense region from 10° south lat. to the Cape Colony in 35° S. feeds immense herds of graminivorous animals the year round, and has done so for ages. The interior of South America is the third great pastoral region; the fourth is Australia; and the fifth is the trans-Missouri and Mississippi country of North America, as yet imperfectly developed, but which perhaps surpasses in every natural advantage any part of the known world.

The number of oxen returned in the United States in 1873 was 16,413,800, valued at £65,860,000, and of milch cows 10,575,000, valued at £63,000,000. The beef product is given there at 2,926,571 tons, of which 2,866,365 tons are consumed, leaving a surplus of 60,206 tons of beef. Of this quantity, about 11,898 tons are shipped, and the remainder, 48,308, unutilised.

The business of stock-raising in the United States is a growing one, particularly in Texas, where extensive experiments are being made with a view to the improvement of cattle by imported stock, and it is believed that ere long, the wild long-horned cattle will be much changed for the better. Illinois ranks second in order, possessing 1,269,000 oxen and 710,900 cows.

The live stock trade of the South-western States has become a matter of millions of cattle. The number in Texas is over 3,500,000, with half as many more upon the western plains. From 350,000 to 500,000 head are annually driven into Kansas,

GROUP OF AFRICAN CATTLE.

Colorado, and Wyoming, to supply the eastern trade. Yearlings are bought at about 21*s.* each, two-year-olds 32*s.* to 40*s.*, cows at 28*s.* to 36*s.*, and oxen at £3; but large herds are often bought much below these prices. A herd of two or three thousand upon the trail presents a fine sight, tramping along in Indian file, extending a distance of a mile or more over the prairie, and feeding upon the spring grasses as they go. At Los Animas, a junction railway station, there are three large slaughter-houses, where 80 or 90 men kill and dress from 700 to 800 oxen a day. They are packed into refrigerator cars, which each hold from 40 to 50 carcases.

Cattle in immense herds are raised in the pastures of Central America. In the plains of Honduras, and on the eastern districts of Nicaragua, there are cattle farms on which are herds of from 10,000 to 40,000 oxen, bulls and cows.

There are four well-marked and distinct genera of the Bovines. 1. Bos. 2. Bison. 3. Buffalo (*Bubalus*), of which there are three small groups: the Buffaloes properly so called, the Arnees, and the Brachycheres; and 4. Ovibos, or Musk Ox.

1. Bos Taurus. The common Ox, or Domestic Cattle.

The breeds of these are almost innumerable, caused by the endless crossings of one breed with another; but as the object here is to treat of their economic uses, rather than the agricultural specialities, it will not be necessary to point out the distinctions into which the different races now recognised have been divided. In Great Britain they are grouped by farmers into "breeds," characterised and named, from various peculiarities, as long-horns, short-horns, polled, &c.

The domestic ox is perhaps the only large animal of whose carcase but one seventh is without some important use.

In many countries, such as India, Africa, and some of the European States, the ox is still yoked to the plough.

In the River Plate district, goods and produce are transported by bullock-carts, which travel about 20 miles a day.

PRIZE SHORT-HORN, "PRIDE OF WINDSOR," SHOWN AT ISLINGTON, 1875.

Cattle are much used for draught by the Dutch farmers of South Africa. Besides the continual transport of produce long distances to market, an instance of the great demand for carriage is afforded in the fact that at the quarterly "nacht maal" or religious gathering of the Dutch at Graham's Town, from 300 to 400 waggons are usually present, each bringing its freight of seven or eight persons, mostly from long distances. A colonial waggon and span of oxen costs from £150 to £250.

Mr. H. Hall says the ox or bullock may be considered, in the absence of railways, as the staple animal for transport purposes all over South Africa. It is used in the vehicles so well known as Cape waggons, in spans of twelve, fourteen, or sixteen, according to the nature of the roads, and will draw from 25 to 40 cwt., although often much more. The rearing and training of oxen forms an important part of Cape farming, and within the last few years the price of a trained ox has risen from £2 10*s.* to £12 or £14. They are generally sold in spans. There are in the Cape Colony above 400,000 draught oxen, besides those in Natal and the Dutch Free States. The pace of an ox waggon averages four miles an hour.

Beyond the frontiers, especially in Namaqua and Damara land, both pack and riding oxen are trained, and very useful animals

AFRICAN BULLOCK-WAGGON.

they are. Oxen depend on a long journey entirely upon what they can pick up in the field.

In Ceylon there are also about 20,000 bullock carts employed.

HEREFORD BULL, "TREDEGAR."

By some naturalists the Cape ox (*Bos Caffer*, Sparrm.) is made a separate species of buffalo (*Bubalus Caffer*, Smith). It has very massive horns, which cannot easily be removed from the core, bending downward and outward, the points swoop upward and inward, their length is about thirty-four inches, span thirty-seven to forty inches.

British Breeds.—The crosses and varieties of English cattle are generally so well known to those who take an interest in stock, that it will only be necessary to enumerate and figure a few of the prominent varieties, giving more attention to descriptions of some of the continental breeds, which are less common.

The Short-horns are an improved breed, now very generally distributed and much esteemed, in favour both with the dairyman and grazier. The colour is red or white, or a mixture of both. All the points of the animal combine to form a symmetrical harmony, which is not surpassed in beauty and sweetness by any other species of the domesticated ox. (See p. 85.)

The Devon is a medium-sized breed, generally of a bright red colour, peculiar to the south of England. They fatten faster, and with less food than most other cattle, and their flesh is excellent. (See Devon Yearling Heifer, p. 119.)

The Hereford oxen are much larger than the Devon, and of a darker red; some are dark yellow, and a few brindled; they generally have white faces, bellies, and throats; they fatten to a much greater weight than the Devons. (See p. 87.)

The Jersey or Alderney is a small delicate breed common in the Channel Islands; they are in general fine boned, and of a light red or yellowish colour. Their milk is very rich, but is rather more yellow or high-coloured than that of other sorts; they are much inclined to fatten, and their beef has a very fine grain, and is well tasted.* (See Jersey Bull, p. 121.)

On the west wall in the Food Gallery there are fine stuffed

* For the illustrations of prize British cattle we are indebted to the *Agricultural Gazette*.

THELEMARK COWS OF NORWAY.

heads of the principal breeds of cattle, such as Shorthorns, Longhorns, polled Angus, Galloway, Hereford, and Scotch neat.

CONTINENTAL BREEDS.

SCANDINAVIAN CATTLE.—The Thelemark race is one of the few constant races of cattle, perhaps the only one, which Norway possesses. It is a well-defined mountain race. The animal is of diminutive size, as is well indicated in the illustration (p. 89) by the stature and attitude of the servant. Full-grown cows rarely attain a greater weight than 660 to 770 lbs. The most remarkable points in the Thelemark breed are the slender form, small head, with long well-shaped horns (on which buttons are usually placed), the sprightly movement, and the bright colouring. This last varies very much, from quite white to tolerably dark, but usually the variations are those of red, spotted, and brindled. More than sixty per cent. of the cattle of Sweden, and more than seventy per cent. of those in Norway, are milch cows. The price of good animals ranges from £5 11*s.* to £6 13*s.* and some few remarkable animals have sold for £11 and upwards.

Most of the country cattle of Sweden are some shade of red, with a certain amount of white, especially about the face, but not so constantly as with our Herefords; the colour also varies from nearly yellow to a deep red. The production of milk and not of meat is the great object of the Scandinavian farmer.

The best type is known as the Herrgards or nobles' race (see p. 91), a name which formerly distinguished it from the less-cultivated type that was bred by the peasants. Mr. Jenkins, in his report on the Agriculture of Sweden and Norway, mentions the fact that one cow of this race gave as much as 920 gallons of milk per annum, and others yielding from 575 to 690 gallons have not been uncommon on the royal estates. Peasants' cows do not, however, yield anything like this quantity, from 200 to 300 gallons being a high average.

AUSTRIAN CATTLE.—Naturalists agree in considering the Hungarian ox as the best living representative of one at least of the original progenitors of our domestic cattle.

These it is believed owe their origin to three distinct types, viz.: *Bos primigenius*, *B. longifrons*, and *B. frontosus*. The two last are extinct as wild races, and are solely represented by certain

COW OF THE SWEDISH HERRGARDS, OR NOBLES' RACE.

types of domesticated cattle. *Bos primigenius* still exists in a semi-wild state in Chillingham Park, and is closely allied to both the Pembroke cattle of South Wales, and the beautiful Devons. The Hungarian and Podolian oxen are also considered to be more or less pure representatives of the *primigenius* type.

The Hungarians are justly proud of their oxen, which are used as working cattle over the whole empire. Professor Wrightson states that it is no uncommon sight to see a team of oxen yoked to a plough, and driven by the ploughman entirely by the voice, without any assistance either from reins or driver.

The Hungarian ox becomes the best and most durable draught ox in the world, remaining useful throughout a long series of years, to be sent at last to slaughter.

The young ox is broken-in in his fourth year, and experience has shown that there is hardly a breed of horses in the world that can compete with the Hungarian ox as a means to agriculture.

PODOLIAN COW, GALICIA.

The Podolian is an aboriginal race of cattle descended from the wild Urus (*Bos primigenius*, Bojanus). Their colour is generally white or silver grey, with variations passing into dark grey, which shade is particularly seen in the bulls. Podolian oxen are much sought after for fattening purposes. Nearly seventy-five per cent. of the oxen slaughtered at Vienna belong to this race. The meat is very much esteemed, and is distinguished for its tenderness and agreeable flavour. It is as working oxen that these cattle are most valued. They will travel two-and-a-quarter miles per hour when yoked to an empty waggon, and one-and-a-half mile per hour when drawing a load. This race is distributed over the greater part of Galicia. The Podolian draught oxen have valuable qualities which render them exceedingly useful in the wide stretching plains which constitute their home.

The Murzthal race is especially prized for its milk-giving properties, and its suitability for draught purposes. The cows have been known to produce 775 gallons per annum, and 464 gallons is

MURZTHAL COW, STYRIA.

given as a usual average. These beautiful cattle are natives of Austria; they are considered to be closely allied to the Hungarian cattle. The oxen work from three to eight years, after which they are fatted. The colour of the hair is badger grey, with brighter stripes round the muzzle; colour rings round the eyes, and dark coloured belly.

The white Norrisch race is known in Styria as the Mariahofer breed. From these and the Murzthalers (another Styrian breed) excellent fat oxen can be obtained. They are distinguished by their small horns, small well-bred head, bright eyes, handsome neck, broad hips, great depth of barrel, short legs, and fine skin. The Murzthalers have the advantage over the Mariahofers in having smaller horns, and being lighter in bone, but the latter are more powerfully developed.

The Montafuner race is principally found in the Montafun Valley, but also in the Bregenzer Forest, and in the Bavarian Allgau.

MONTAFUN COW, VORARLBERG—TYROLESE RACE.

The Mariahof cattle are natives of Western Styria, are noted as milking cattle, and also fatten readily. They are a uniformly coloured race, often white, but sometimes inclining to lighter and darker degrees of fawn. They number nearly 300,000 head. (See cut, p. 112.)

This tribe, which is connected with the Swiss cattle, belongs to the heavy average group. The cows reach a live weight of from 8 to 9 cwt., being lighter than the Swiss and heavier than the Allgauer.

The colour does not generally differ from that of the Swiss race brown and brownish grey colours are predominant. These animals are highly distinguished for their good temper, and their capabilities for draught and fattening purposes are satisfactory.

The Egerland cattle resemble the reddish brown Tyrolese race in their general characters, and are said to be the result of the crossing of the Bohemian native race with Zillerthal bulls. They have small heads, and are of a nearly uniform dark reddish brown colour. The average weight of a full-grown ox may reach as much as 9½ cwt., and the animal will fetch from £20 to £22 10*s*.

The Egerlander prides himself upon the beautiful and regularly formed horns of his oxen, and he assists to develop them whenever their growth is not naturally uniform. This is done by weights, which are connected by cords to rollers attached to the roof of the stables; these weights thus follow all the movements of the heads of the cattle; clamps are also used, by means of which the horns, after having been softened with grease, are pressed either forwards, backwards, or sideways, as may be required. (See cut, p. 107.)

The animals of the Pinzgau breed are greatly valued, not only for their milk-giving properties, but on account of their rapid development. (See cut, p. 109.)

This race is the result of a cross of the Simmenthal breed with the ancient domestic cattle of the country; it is distributed nearly throughout the whole of the Salzburg region, and in certain parts of the Tyrol adjacent, in Upper Austria, and parts of Bavaria. It has a very symmetrical and pleasing appearance, and is of a red colour, varying from light red to brownish red, with many white patches. It is a very desirable butcher's race.

The so-called "Kuhland" district, which takes its name from its suitability for cattle-breeding, is situated in north-western Moravia, on the flanks of the Carpathians. The predominant colour of these cattle is dappled red or cherry red, with large white patches on the head, along the back, and on the belly; they have also the finest, softest, silky hair, and the forehead bears a strong tuft of crisped or curled hair. (See cut, p. 117.)

Although of middle height the Kuhland cattle must be classed with the heavier races. The live weight of a cow may be taken at from 6 to 10 cwt.; the cows yield $1\frac{1}{2}$ to 3 gallons of milk per day.*

The Mysore ox is deep in the chest, roomy in barrel, fine in the legs, straight-horned, and sleek of coat. The prevailing hue is a light cream colour. It is the carriage ox in India, the Arab among bullocks.

The Indian or Brahmin bull, often called the zebu (*Bos indicus*), extends over Southern Asia and the Eastern Islands, and is found

BRAHMIN BULL.

also in Eastern Africa. It is common in the north-west Himalayas, but rarely seen in the Nepaul mountains. They are venerated by the Hindus, who object to slaughter them, but use them in harness, and they will travel about thirty miles a day. These oxen have pendulous ears, and are distinguished by a fatty elevated hump

* For the various illustrations of Continental breeds of cattle we are indebted to the "Journal of the Royal Agricultural Society of England."

upon the withers, which sometimes weighs 50 lbs., and, when properly cooked, is said to be delicious. The flesh of the animal is not, however, so palatable as that of the common ox.

On the west wall are five or six fine heads and horns of the Gaur (*Bos Gaurus*), the head of an Indian cow, fine samples of buffalo horns, and a curious horn curled by disease.

2. Bison.—The Bisons (*Bos Bison—Bison Americanus*) are easily distinguished by their highly developed hump, giving them an extraordinary height at the withers, and also by the long hair which covers the anterior portions of their bodies.

In North America the name of *buffalo* has been universally, but inaccurately, given to the bison, and we are hence constrained to speak of it frequently under this misnomer. They bear about as much resemblance to an Eastern buffalo as they do to the zebra or common ox. The general colour of this animal is a uniform dark brownish dun. The Cape ox is also often spoken of as the buffalo.

Imagination can scarcely realise the numbers of bison which even now are found on the western plains. It is not uncommon to see the prairies covered with them as far as the eye can reach, and travellers have passed through them for days and days in succession, with scarcely any apparent diminution in the mass. Explorers have often given almost incredible accounts of the numbers met with. The late Horace Greeley, writing from the plains, says, "I know a million is a great many, but I am confident we saw that number yesterday. Certainly all we saw could not stand on ten square miles of ground. Often the country for miles on either hand seemed quite black with them * * * * Consider that we have traversed more than one hundred miles in width since we first struck them, and that for most of this distance the buffalo has been constantly in sight, and that they continue for some twenty miles farther on—this being the breadth of their present range, which has a length of perhaps a thousand miles, and you have some approach to an idea of their countless millions.

I doubt whether the domesticated horned cattle of the United States equal the numbers, while they must fall considerably short in weight of the wild ones."

The Exploring Expedition of Governor Stevens on the northern route of the Pacific railroad, was frequently arrested for a considerable time by herds of bison, amounting, in some instances, to not less than half a million each. As the expedition rose over the verge of some elevation in the prairie, before them, as far as the eye could reach, stretched an apparently interminable sea of flesh.

It would be supposed that the immense slaughter continuously carried on would have decimated these animals; but, according to recent accounts given by Major Twining in the "Smithsonian Reports," the immense herds are constantly increasing in the north-western Montana; they merely shift their grazing ground. Three hundred thousand human beings depend for their very lives and for everything—according to their savage notions—worth living for, solely and entirely on the bison. Its flesh is their only meat, and most of them will go a long time hungry rather than eat small game or wildfowl. The skin serves them for coats, beds, and boots, walls for their tents, and tiles for the roof—for saddles, bridles, and lassoes. The bones are converted into saddle-trees, into war clubs, whistles, and other musical instruments. Seven trains of railway cars, freighted with buffalo bones, recently arrived in New York to be worked up into button moulds, knife handles, and other uses. Of the horns are made ladles, and spoons, and pins; the sinews serve for strings to bows, and for the attachment to their persons of scalps and such other articles of vertu as fall in the native's way. The hoofs and horns when stewed yield a superior glue, which is largely used in the construction of hunting spears and arrows. The hair of the mane is twisted into ropes and horse halters; the brains even are not wasted, but used in the preparation of buffalo robes, leather thongs, and other articles made from the hide.

The uses of the bison when dead are various. Powder flasks are made of their horns, and they are used for mounting knives and awls. The skin forms an excellent buff leather, and when dressed with the hair on, serves the Indians for clothes and shoes. "Buffalo robes" are generally the dried and prepared skins divided into two parts. A strip is taken from each half on the back of the skin where the hump was, and the two halves or sides are sewed together with thread made of the sinews of the animal,

BOS AMERICANUS (POPULARLY MISNAMED THE BUFFALO IN AMERICA).

and, after much dressing, the robe is ready for market. The number brought into commerce varies, according to the demand, from 60,000 to 100,000, annually. They are much used by the Indians themselves for blankets, clothing, and constructing their lodges or tents. Not above a tenth of those slaughtered furnish a sufficient furry coat to serve as a robe.

The buffalo robe has of late years been as much used in Europe as it is in North America for a warm travelling wrapper; it sells at from £3 to £10, or even higher; and in the cold climates

H 2

of Europe it is similarly employed for sleigh wrappers, cloak and coat linings, &c.

The flesh is a considerable article of food, and the hunch on the shoulders is esteemed a great delicacy. One of the most useful applications of buffalo meat consists in the preparation of pemmican, an article of food of the greatest importance in a northern climate from its portability and nutritious qualities. This is prepared by cutting the lean meat into thin slices, exposing it to the heat of the sun or fire, and when dry pounding it to a powder. It is then mixed with an equal weight of buffalo suet, and stuffed into bladders. Each bison will produce from 50 to 70 lbs. of tallow, but a bull bison, when fat, will frequently yield 150 lbs. weight of tallow, which forms a considerable article of commerce. The hair or wool is spun into gloves, stockings, and gaiters, that are very strong, and look as well as those made of the finest sheep's wool.* The tail, mounted on a wooden stand, ornamented with goose or porcupine quills, is used as a whisk or fly-flapper.

An attempt is being made by a farmer of Massachusetts to domesticate the bison, and having transported several of these wild roamers of the prairies to his stock farm, he intends to try a cross with Jersey, Ayrshire, or Durham cattle.

It is killed in immense numbers by the North American Indians, solely for the tongue, the skin, and the bosses or humps. They have a peculiar method of dressing the skin with the brains of the animal, in which state it is always imported.

3. Buffalo (*Bubalus*).—The name buffalo is scientifically restricted to a species of ox found in various parts of India and the Eastern islands, and to a more limited extent in some parts of Europe and Africa. The buffalo is of Indian origin, and was only brought into Italy a little before the sixth century, but has spread over South-eastern Europe and the North of Africa.

* In Case **58** will be found black and grey worsted yarns, spun from the wool of the bison.

In 1870 there were 73,153 head of buffalo in the Hungarian dominion; in Greece, Piedmont, Italy, and Spain, the buffalo is also found and esteemed as an animal of draught. But he must have water to revel in, and hence thrives only in fenny land. The accompanying is a sketch of a European buffalo bull, thus described by Prof. Wrightson: "The colour is completely black, hair and skin, hoofs and horns, all partaking of this sable hue.

EUROPEAN BUFFALO BULL.

The limbs are short and thick, the body massive, the head large, the forehead arched and narrow, the muzzle large and black, horns low placed, triangular at base, furrowed across and directed backwards and downwards, finally turning upwards towards the point. The hair is scattered somewhat thinly over the body of the full-grown animal, although the calves are well covered."

Buffalo milk is an ingredient in the Transylvanian national diet which cannot be dispensed with upon great occasions. It is richer than that yielded by any other animal. In South Hungary and Transylvania no gentleman considers his breakfast complete without buffalo milk with his coffee. The largest milkers give 6 quarts per day, but 2 or 3 quarts is a more ordinary quantity. The flesh is "stringy," and gives off a musky odour which spreads all over the house, and also affects the milk. For this reason buffalo beef is seldom used, although the veal is considered good. The skin makes good leather.

The European buffaloes which were shown at the Vienna Exhibition in 1873, attracted interest from their peculiar smell, as well as shape. The cows are remarkable for their breadth of hip (crux) which is a peculiarity of the buffalo.

Of the number of these animals existing in Eastern countries we have no certain data. According to a late census there were, however, in British Burmah, 600,000 buffaloes, and 566,000 cows and oxen. In Java, in 1873, there were 2,750,000, besides 1,628,000 other cattle.

The Indian buffaloes (*Bubalus buflus*) are of large size, but low in proportion to their bulk.

The buffalo is poorly fed, not generally cared for, and usually killed when too old to breed or give milk. The hide of the male buffalo is coarse, and it gets such bad treatment in the plough or cart that it is generally full of sores and goad marks. In large towns there is a market for buffalo beef for the low caste and poorer Mussulman population, and also for grease, and younger and better cattle are slaughtered; it is from these that the local tanners select their hides for the finer uses of harness, saddlery, and accoutrements.

A pigment known as "purree" or Indian yellow is produced in Monghyr from the urine of horned cattle fed on decayed and yellow mango leaves. It is used in the locality of production, and also sent to Calcutta for export.*

The buffalo possesses two excellent qualities; he is immensely strong, and his wants are easily satisfied. The buffalo is the beast of burden in the Indian archipelago and parts of Asia. Being

* A specimen of this will be found in the Case of Animal Dyes, No. 162.

thickset in form, with large members and powerful muscles, two draught animals will draw as much as four ordinary horses or six oxen. The flesh is not palatable, and the milk of the female is less pleasant than that of the cow, having a musky flavour, but as it is abundant, it is sometimes made into cheese.

From the great size of the buffaloes and most of the oxen in Java, the hides and horns are peculiarly valuable. The immense horns of the Java buffalo have long formed an article of European trade, and the hides are sent to China in the hair and untanned. Bali and Lombok supply a great number. The hides available for export in Java, are greatly diminished by the singular practice among the inhabitants of that island of using the fresh hide as an article of food, nay even esteeming it a dainty beyond any other part of the animal. The increased demand for hides as an article of commerce has, however, somewhat tended to put an end to this taste.

In Sumatra they dress their meat immediately after killing it, while it is still warm, which is conformable with the practice of the ancients, as recorded in Homer and elsewhere, and in this state it is said to eat tenderer than when kept for a day; longer the climate will not admit of, unless when it is preserved in that mode called "dendeng." This is the flesh of the buffalo cut into small thin steaks and exposed to the heat of the sun in fine weather, generally on the thatch of their houses, till it has become so dry and hard as to resist putrefaction without any assistance from salt. It is seemingly strange, that heat which, in a certain degree promotes putrefaction, should, when violently increased, operate to prevent it. A large export trade is carried on in this dried meat from the islands of the Eastern Archipelago to China and Siam.

The Arnee buffalo (*Bos Arnee*, Shaw—*Bubalus Arnee*, Smith) has horns of a prodigious size and length, which are turned laterally, flattened in front and wrinkled on the concave surface.

The Yak or grunting ox (*Poephagus grunniens*), is often

handsome and a true bison in appearance. It is met with wild in Central Asia, and is the largest native animal of Thibet, in various parts of which country it is found Domesticated the yak is used as a beast of burden. It is invaluable to the mountaineers of Thibet and other parts of Central Asia from its strength and hardiness, accomplishing, at a slow pace, twenty miles a day, bearing either two bags of salt or rice, or four to six planks of pine wood slung in pairs on either flank. It is ridden especially by the fat Lamas, who find its shaggy coat warm, and its paces easy. Under these circumstances it is always led. They have spreading horns, long silky black hair and grand bushy tails; black is their prevailing colour, but red, dun, parti-coloured, and white are common. The hair is spun into ropes, and woven into a covering for their tents; the gauze shades for the eyes used in crossing snowy passes are made from the same material.*

The flesh of the young yak, according to Dr. Hooker, is delicious, much richer and more juicy than common veal; that of the old yak is sliced and dried in the sun, forming jerked meat, which is eaten raw, the scanty proportion of fat preventing its becoming very rancid. Opinions differ as to the quality of the meat. Pallas says the flesh is hard and bad tasted; Huc, on the contrary, asserts it to be very good. Much of the wealth of the people consists in its rich milk and curd, eaten either fresh or dried, or powdered into a kind of meal. It forms an important article of commerce, and will keep fresh in skins a long time.

The Lepchas eat the flesh and entrails, and singe and fry the skin of the yak, and make soup of the bones, leaving nothing but the horns and hoofs. They also prepare the flesh as jerked meat, cutting it into strips, which they dry on the rocks. This, called shat-chew (dried meat), is a very common and palatable food in Thibet.

* The hair and tail of the yak are shown in Case 97.

The yak inhabits the southern slope of the Himalayas between the 27th and 28th degrees of N. lat., and extends from there to Little Thibet or Ladak, Grand Thibet, and the north of China, and becomes rare in Mongolia. It is found there, both in the wild and domesticated state. The height of the animal varies.

The yak is to the mountaineers of Thibet, what the horse, the ass, the cow, and the sheep are to Europe. It furnishes food to the inhabitants in its flesh and its milk, it carries heavy burdens

THE YAK, OR GRUNTING OX.

takes to the yoke easy, and serves for agricultural work. From its long and abundant fleece is obtained a silky wool, which serves to make warm and strong clothing. The furry coat of the young animals is curly and woolly, resembling that known as Astracan, obtained from a breed of sheep. Very fecund and hardy, they resist the most rigorous cold and brave intemperate seasons. The inhabitants of Thibet have the same respect for the yak which the Brahmins have for the Zebu.

The bushy tail of the yak, garnished with handsome hair, more fine and soft than that of the horse, is much esteemed in the plains of India as a chowree or fly-flapper, and by the Chinese, who dye it bright red or blue for all sorts of ornaments. Among the Persians, Turks, and some other Eastern nations, the yak's tail is a standard or emblem of authority, and a distinctive mark of certain military dignitaries.

The tail and long hair of the yak occasionally come into commerce in London.

4. The Musk ox (*Ovibos moschatus*), is a small but powerful animal inhabiting the northern regions of the American continent. As it is very scarce and difficult of access, and the flesh is strongly impregnated with the flavour of musk, it is not much looked after. The calf-skins make excellent robes and caps, but the adult hides are almost too hairy for any purpose of that sort; its hair reaches almost to the ground, so that it has rather the appearance of a long-haired goat than of a true ox. The hair is matted, somewhat curled and bushy, and in general of a sombre brown colour. The tails are made into fly-flappers, similar to those obtained from the bison and the yak.

When the animal is fat, its flesh is said to be excellent; but at certain periods, and especially early in the spring, the strong musky odour renders it unsavoury. In winter the long woolly hair is thick and dense, and enables the animal to withstand with impunity the severe frosts. If this wool could be easily obtained, and in sufficient quantity, it would be much in demand for manufacturing uses from its silky character.

Our Meat Supply from Cattle.—The enhanced consumption of butchers' meat here, and the advance in price, have become important considerations, and more especially as to our future adequate supply; the graziers being unable to meet the increasing demand. The larger consumption arises from two causes, which, in all probability, will continue to operate, namely, the increase of our population, at the rate of 1000 per day, and the improvement in the social condition of the operative classes, consequent upon the enormous extension of commerce and manufactures, and the abundant employment created by railways, building operations, and other public works.

For the daily supply of the metropolis about 4,500 live beasts are sent to the London market; 300 tons of dead meat come by railway from the North; and the amount sold daily in the Metropolitan Dead Meat Market alone is 500 tons.

EGERLAND COW, BOHEMIA.

Official returns showed that the annual consumption of meat in the kingdom in 1870 was about 96 lbs. per head; this, at the wholesale price of 8*d.* per pound, amounted to 64*s.* for each person, and for the entire population about £99,000,000.

Mr. H. S. Thompson in the "Journal of the Royal Agricultural Society," recently estimated the average consumption of meat by each person in the United Kingdom at 103 lbs. per annum; for in 1871, 31,700,000 people consumed about 1,500,000 tons. This is equal to a demand for 1,300 head of cattle, sheep, and pigs every day. In 1872 the average consumption of meat had risen to 108 lbs. per head. Upwards of £220,000,000 is said to be invested in live stock in Great Britain and Ireland.

The number of cattle and calves in the United Kingdom in 1875 was 10,162,787. Of these, two-sevenths, or about 3,000,000, are supposed to be annually slaughtered, the average weight per head all round (for both cattle and calves) being taken at 560 lbs. To this has to be added the supply of mutton and pork and other animal food, fresh or cured.

Our annual import of beef, salted or fresh, is about 250,000 cwts.; besides which there is a further quantity of undescribed meat imported, as shown by the following figures, in cwts.:

	1860.	1870.	1875.
Meat—Salted or fresh	15,007	84,300	144,987
,, Preserved	6,131	80,636	171,746

The foreign food supplies, chiefly derived from horned cattle, received in 1875, were: salt beef, 181,504 cwts.; beef fresh or slightly salted, 35,000 cwts.; butter, 1,467,000 cwts.; and cheese, 1,626,413 cwts.

According to the experiments of Messrs. Lawes and Gilbert, the following is the assumed average per-centage composition of the entire carcases of butchers' meat as fattened for slaughter, determined by chemical analyses:

	Mineral Matters.	Dry Nitrogenous Substances.	Fat.	Total Dry Substances.	Water.
Calf	4·5	16·5	16·5	37·5	62·5
Bullock	5·0	15·0	30·0	50·0	50·0
Lamb	3·5	11·0	35·0	49·5	50·5
Sheep	3·5	12·5	40·0	56·0	44·0

The actual weight and composition of the carcases in pounds is:

	Total Weight.	Mineral Matters.	Dry Nitrogenous Substances.	Fat.	Total Dry Substances.	Water.
Calf	150	6¾	24¾	24¾	56¼	93¾
Bullock	900	45	135	270	450	450
Lamb...............	45	1½	5	15¾	22¼	22¾
Sheep	90	3⅕	11¼	36	50⅖	39⅗

Later investigations by Messrs. Leyder and Pero show that during the process of fattening animals the quantity of dry material is notably increased, for while in oxen in moderately poor condition the water is about two-thirds of its total weight, in a fat ox it is only a half. The more nutritious character and superior taste of the flesh of a fat animal are due to this increase

PINZGAU COW, SALZBERG RACE.

of dry material, but of this increase two-thirds consist in fat; the increase of proteids is only from 7 per cent. to 8 per cent., and of inorganic materials 1½ per cent. From a variety of analyses the flesh of the fat animal is in every case found richer in fixed material than that of the lean animal; and though the flesh of a lean animal possesses a more uniform quality than that of a fat one, yet the poorest parts in the fat one possess a higher nourishing value than the best in the lean animal.

The wholesale price of prime beef per stone of 8lb. in the Metropolitan Market has been as follows:

		s.	d.
Average price for five years ending 1853	...	4	$2\frac{1}{2}$
,, ,, 1863		5	$0\frac{1}{2}$
,, ,, 1873	...	5	$6\frac{1}{2}$
,, Two years, 1874 and 1875		5	$8\frac{1}{2}$

so that there has been an increase of 18*d.* in price, or an advance of $35\frac{1}{2}$ per cent., in 22 years.

The mode of cutting up meat varies in different towns and countries. In the carcase of every animal, an ox, for instance,

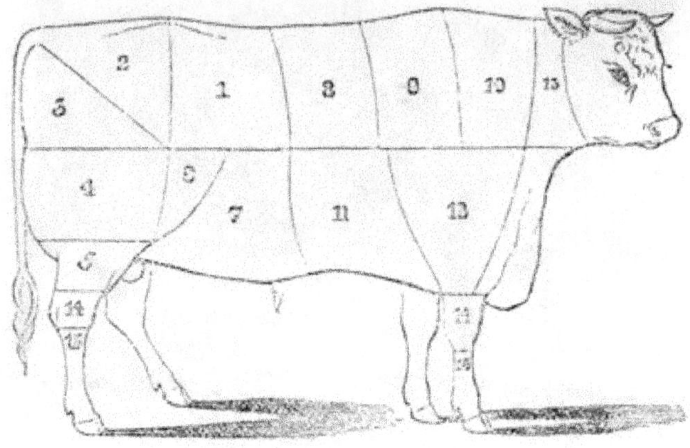

there are various qualities of meat, and these are situated in different parts of the carcase. All the best parts are in London used for roasting and steaks, and the inferior for boiling, or making soups, &c. The precision with which expert London butchers divide the different qualities of meat from the same carcase shows their thorough knowledge of the kinds, and the tastes of the grades of customers they have to supply.

The carcase of an ox is in London cut up into the following pieces, as may be seen on referring to the numbers on the above cut.

LONDON MODE OF DIVIDING A CARCASE.

HIND QUARTER.	FORE QUARTER.
1. Loin.	8. Fore-rib.
2. Rump.	9. Middle-rib.
3. Itch, or adze-bone.	10. Chuck-rib.
4. Buttock.	11. Brisket.
5. Hock.	12. Leg of mutton piece.
6. Thick flank.	13. Clod and sticking and neck.
7. Thin flank.	14. Shin.
	15. Leg.

In Paris the butchers estimate the proportion of flesh to the weight of the live animal at about 58 per cent. The various parts of an animal weighing 600 kilogrammes (1320 lbs.), according to their rule, would be as follows:*

	KILOS.
Meat	350
Hide	50
Fat	40
Blood	25
Feet and hoofs	10
Head (bones and brains)	5
Tongue	3
Liver and milt	10
Lungs and heart	7
Intestines	30
Waste, evaporation, &c.	70
	600

Our importation of foreign live cattle is confined to a few neighbouring States, owing to the danger and difficulty of longer sea voyages, which involve a heavy per-centage of loss by deaths.

From those countries, too, whence we obtain the largest and best supply of cattle and sheep, the exportation appears to have reached its maximum. It was in 1842 that we commenced to import live animals, when the number of cattle received from abroad was but 4,264. In 1853 the number of cattle received was 125,253, and of sheep 259,420. It then declined somewhat until 1863, when the cattle imported numbered 150,898 animals,

* "Des Halles et Marchés," par J. R. De Massy. Paris, 1862.

and in the three following years averaged about 250,000 head. Since then there have been fluctuations owing to the restraints arising from examination for disease on landing. In 1875, 263,698 head of foreign cattle were imported.

MARIAHOF COW, STYRIA.

According to an official report published in 1862, the average consumption of butcher's meat in Paris (exclusive of pork) was as much as 140 lbs. per head.

In New York the average consumption of meat per head is said to be as high as half a pound per day, which would be at the rate of 182 lbs. per annum. But even this is exceeded by the consumption in Buenos Ayres, Uruguay, &c., where it is stated to average a pound of meat per day.

The meat production of Russia is not very definitely fixed, but an official publication * estimated in 1867 that 3,500,000 head of large cattle were annually slaughtered, of which 2,200,000 were oxen, and 1,300,000 cows. The number of calves killed was

* "Aperçu Statistique des Forces Productives de la Russie," par M. de Buschen. Paris, 1867.

GROUP OF CATTLE.

stated at 4,000,000, and of sheep 12,000,000. Allowing, from authentic data, each head of cattle to average 450 lbs. of meat and 60 lbs. of tallow (an ox yields 550 lbs. of meat and 100 lbs. of tallow, and a cow 250 to 300 lbs. of meat and 150* lbs. of tallow), we have a total of about 703,000 tons of meat, and about 93,750 tons of tallow. The four million calves at 80 lbs. each would give 142,850 tons of meat. Reckoning the sheep at 30 or 40 lbs. of mutton each, and 10 lbs. of tallow, we have a further quantity of 168,700 tons of meat, and 53,570 tons of tallow.

The average consumption of meat per head in Russia therefore cannot be estimated higher than 40 lbs., exclusive of pork.

In Spain the meat consumption is only 50 lbs. per head in the large towns, and half that in the smaller towns; and yet the country possesses large quantities of food-producing animals.

DRIED MEAT.—There is a large trade carried on in Brazil and the River Plate States in charqui or jerked beef. The exports from Buenos Ayres alone average 600,000 to 700,000 cwts. yearly. The labouring classes in nearly every part of South America and the West Indies live almost exclusively upon this jerked beef, which is prepared by cutting the meat into ribbon-like pieces, and drying them in the sun with a small addition of salt, or by steeping them in a strong pickle for twenty-four hours.

Most of the saladeros, or slaughter places, supply their wants directly from the estancias or cattle farms, whence the animals are driven distances varying from twenty to fifty leagues, and often farther. The transit of the animals is by no means an easy task, as they are generally very wild. In the camp, when night comes on, they have to be rounded and watched by several men, and it often occurs, particularly in wet weather, accompanied by thunder and lightning, that a "stampede" takes place, the animals breaking loose and giving no end of trouble to be caught again. It therefore not unfrequently happens that there are several animals missing before the troop arrives

* This must be a misprint in the official work for 50 lbs.

at its destination, for which risk and work the party who contracts to deliver them is paid about 8*s.* per animal. The first cost ranges from £2 8*s.* to £3 4*s.* per head, and the total expenses on the beast from the time it is placed in the saladero until its produce is shipped, including lighterage and export duty, 16*s.* per head.

On arrival at the saladero the animals are driven into enclosures, called "corrales," and are slaughtered as soon as possible.

In Brazil the heat of the climate precludes the salting of beef in large pieces or joints. The province of Rio Grande do Sul, which enjoys a most temperate climate, is the seat of the charqueadas or slaughtering establishments, and produces great quantities of the "carnas do sertaon," or meats for the interior, sufficient for home consumption, and even for exportation. Each animal yields about one hundredweight and a half of dried beef; 300,000 to 400,000 cwts. are sent annually to Cuba for the black population. Chili ships about 4,000 to 5,000 cwts. of this charqui or sun-dried beef.

This custom of drying meat in the sun is, as we have seen, carried on in the East, and also practised in Africa, for Capt. Burton, in his "Lake Regions of Central Africa," states:

"The African preserves his meat by placing large lumps upon a little platform of green reeds, erected upon uprights about 18 inches high, and by smoking it with a slow fire. Thus prepared, and with the addition of a little salt, the provision will last for several days, and the porters will not object to increase their loads by three or four pounds of the article, disposed upon a long stick like gigantic kababs. They also jerk their stores by exposing the meat upon a rope, or spread upon a flat stone, for two or three days in the sun; it loses a considerable portion of nutriment, but it packs into a conveniently small compass. This jerked meat, when dried, broken into small pieces, and stored in gourds, or in pots full of clarified and melted butter, forms the celebrated travelling provision in the East called kavurmek: it is eaten as a relish with rice and other boiled grains."

Dairy Products.—Having dealt with the flesh of cattle used as food, we come now to consider briefly the MILK, BUTTER, and CHEESE, which are furnished by cows.

Milk.—In various countries different animals have been used to provide milk for the use of man; in fact, it seems probable that every domestic animal except those which are carnivorous has been put to this service. Besides the cow, from which our principal supplies are provided, the following animals are used in their respective countries. The goat in most mountainous regions, to some extent in England, and in parts of Switzerland; the sheep in several pastoral countries, and in former times occasionally in England; the reindeer in Lapland; the camel by the Bedouins and others who use this animal; the sow in China; the mare by many Central Asian tribes; the ass commonly in various countries.

The Laplanders preserve the milk of the reindeer in frozen pieces like cheese. When melted, after a lapse of several months, it still tastes fresh and good. When a stranger enters their dwelling whom they wish to welcome, the frozen piece of milk is immediately set to the fire; the guest receives a spoon, with which he skims off the softened exterior as it melts; when he has had enough the rest is preserved in the cold for other guests.

The milk of the buffalo is like that of the cow. Goats' milk stands next to these in its qualities, and is much used in Southern Europe. The milk of the ewe is richer than that of the cow in fat, and contains rather more sugar than that of other animals. The Calmucks, and most of the tribes of Central Asia, prepare a beverage from camel's and mare's milk, which is also fermented and distilled into an alcoholic beverage. This koumis, as it is termed, has lately been recommended medicinally in this country. The milk of the mare is inferior in oily matter to that of the cow, but contains a fair proportion of sugar and other salts. The milk of the ass approaches that of human milk in several of its qualities, and is recommended for invalids in pulmonary complaints. Camel's milk is poor in every respect, but is employed in countries where the animal flourishes. Every preparation of milk, and every separate ingredient of it, is wholesome.

Milk, Captain Burton tells us, is held in high esteem by all the tribes of Central Africa. It is consumed in three forms—fresh; converted into butter-milk; and in the shape of curded

KUHLAND COW, MORAVIA.

milk. The latter is everywhere a favourite on account of its thirst-quenching properties, and the people accustomed to it from infancy have for it an excessive longing. It is procurable in every village where cows are kept, whereas that newly drawn is generally half soured from being at once stored in the earthen pots used for curding it. Buttermilk is procurable only in those parts of the country where the people have an abundance of cattle.

The aggregate consumption of milk in the United Kingdom is very large, and may be roughly estimated at a quart a week for each person. At this rate 812,500 gallons would be required for the weekly supply of London alone, with its population of 3,250,000. The average yield of milch kine is variously estimated, ranging between 2 quarts and 20 quarts a day; assume 8 quarts as a fair average, about 406,250 cows are required to furnish the metropolitan supply alone, and the consumers must pay more than 3½ millions sterling per annum for milk in London.

In Paris, even with its smaller population, the aggregate consumption of milk is larger than in London. In 1860 it was returned at over 300,000 quarts daily.

Milk, in the language of the dairyman, is composed of three substances—butter, cheese, and whey; and to separate the two former from the latter is one of the chief occupations of the dairy. A quart of milk of fair average quality should weigh 2 lbs. 2¼ ounces.

BUTTER.—The usual allowance of butter to domestic servants is half a pound per week, but if we assume a consumption of only 20 lbs. yearly, for each individual of two-thirds of the population, this would require a total supply of 200,000 tons a year, and our home production must be fully 130,000 tons. We import now about 80,000 tons of foreign butter, for which we pay more than £8,500,000 sterling. In 1858 the consumption of foreign butter was only 1·52 lb. per head, now it is as much as 5·51 lbs. per head of the population. The quantity received from abroad has not varied much of late years; the average imports are about 1½ million cwts. Foreign butters from Holland and France are preferred to the Irish butter, because they are so fresh and scrupulously clean; Irish butter often contains hair and dirt of various kinds, as well as too much salt and brine. Irish butter is sub-divided into six qualities or classes.

In Paris the consumption of butter in 1850 was only about 18,000,000 lbs.; in 1860 it had increased to 30,400,000 lbs., being an average of 25 lbs. per head per annum for the population. In the first rank stands the butter of Isigny, which includes not only

DEVON YEARLING HEIFER, SHOWN AT CROYDON, 1875.

the butter made in the locality of the Department of La Manche, from which it takes its name, but also the superior butters of Normandy and Calvados. After this comes the Gournay butter, made in the departments of Eure and Seine-Inférieure. The salted butter comes from Brittany, especially Morlaix, Rennes, Nantes, and Vannes.

The quality of butter, either as regards its keeping properties or otherwise, is affected by the weather, by the condition of the milk, the description of cattle, by the pasture, by the size, airiness and convenience of the dairies, and very much by the sort of fuel used in the district; for where peat or turf is burned, the butter generally takes a flavour from it. Butter intended for keeping ought to be thoroughly freed from the milk in making, the cream being in good condition, and not injured by heat, and the butter should be made close in grain, firm, and not too rich. Such butter does not require the great quantity of salt that is necessary for butter not possessing those keeping properties.

Munster is the great butter-producing province of Ireland, Cork being the seat and centre of the trade, especially the foreign, which from many causes, it entirely monopolises. Waterford is a great butter shipping port, and Limerick and Belfast also supply considerable and increasing quantities for the English market. The butter of the above three towns is more adapted than that of Cork for immediate consumption from its lighter cure, less salt being used in its preparation; whereas the Cork butter, being more heavily salted, can be preserved much longer than the others.

The receipts of butter in the chief Irish markets have steadily increased of late years, at least one-third being destined for foreign consumption. The principal foreign trade is with Brazil, Portugal, the West Indies, and the Mediterranean, and large shipments are also made to Melbourne. The great Irish foreign trade in butter is now with Brazil.

The average amount of butter from milk in summer is rather

JERSEY BULL, "GIPSY KING."

more than an ounce to a quart, or from 16 quarts of milk to 17 to 18 ozs. of butter.

The yield of butter from a good cow ought to be a pound per day, and double this quantity has been produced for a limited period. In Holland each cow, after being some time at grass, yields about one Dutch pound (17½ ozs.) per day.

There are three distinct kinds of butter manufactured in Holland. The butter made from the cream when the cows are at the grass in summer, called *grass butter;* the butter from the whey of the sweet-milk cheese, called *whey butter;* and the butter made in winter, when the cows are in the byre, called *hay butter.* Grass butter is made in the following manner. The cows being carefully milked to the last drop, the copper pitchers, lined with brass, or pitchers entirely of brass, which contain the milk, are put into an oblong water-tight pit, called a *koelback*, built of brick or stone, about six feet in length, three in breadth, and two in depth, into which cold water has been previously pumped, there being generally a pump at one end. In this pit or cooler the pitchers stand two hours, the milk being frequently stirred. This cooling process is of great advantage in causing the cream to separate rapidly and abundantly from the milk. The milk is then run through horse-hair searchers or strainers, and is put into flat milk dishes of earthenware, copper, or wood. After remaining in a cool milk-house or cellar for twenty-four hours, it is skimmed, and the cream is collected in a tub or barrel. When soured, and if there is a sufficient quantity from the number of cows, they churn every twenty-four hours, the churn being half-filled with the sour cream. A little warm water, near the boiling point, is added in winter to give the whole the proper degree of heat, and in very warm weather the cream is first cooled in the *koelback*, or cooler. In many small farm houses, or when the cows give little milk, the milk is not skimmed, but the whole when soured is put into the churn; the butter, immediately after being taken out, is put into a shallow tub called a *vloot*, and then carefully washed with pure cold water. It is then worked with a light sprinkling of finely ground salt, whether for immediate use or for the barrel; there being none made entirely fresh or without salt, as in Scotland and England. If intended for barrelling, the butter is worked up, twice or thrice a day, with soft fine salt, for three days in a flat tub, there being about two pounds of this salt allowed to fourteen pounds of butter; the butter is then hard packed by thin layers into the casks, which casks are previously carefully seasoned and cleaned. They are always of oak, well-smoothed inside. Before being used, they are allowed to stand three or four days filled with sour whey, and afterwards carefully washed out and dried. Hay butter undergoes the same process, being of course the butter made in winter, but although inferior in flavour and colour, it has none of the disagreeable taste which the turnip imparts to much of the winter butter of this country. Whey butter is made from the whey of the sweet-milk cheeses. The whey, being collected from the curd and the pressed cheese, is allowed to stand three days or a week according to the quantity; the cream is either skimmed off or churned, or the whey itself is put into churn, and the butter is formed in about half an hour. In winter the butter obtained by the process is about 1 lb. per cow per week, and in summer about 1½ lb. per cow per week.

Without giving credence to all the exaggerated statements of the quantity of artificial butters foisted on the market, there is too much reason to believe that melted tallow largely takes the place of genuine butter in the retail trade. Very many factories for preparing artificial butter have been started on the Continent and in America. When people come to know that rancid butter is as certainly poisonous as rancid tallow, they will be more cautious about eating it.

If a mass of crude animal fat be heated to a temperature not exceeding 120° Fah. the whole of it will melt, and the product be perfectly odourless, and available for domestic and cooking purposes. Microscopic examination with polarised light is the most reliable means of distinguishing pure butter from that which contains an admixture of other less palatable fats.

To render butter capable of being kept for any length of time in a fresh condition, that is, as a pure solid oil, all that is necessary is to boil it in a pan till the water is removed, which is marked by the cessation of violent ebullition. By allowing the liquid oil to stand for a little, the curd subsides, and the oil may then be poured off, or it may be strained through calico or muslin into a bottle and corked up. When it is to be used it may be gently heated and poured out of the bottle, or cut out by means of a knife or cheese gouge. This is the usual method of preserving butter on the Continent, and also ghee in India.

Ghee is the clarified butter of Hindostan. It is produced generally from the milk of buffaloes, and is universally used in native cookery. As an article of commerce ghee possesses some claims to importance, many thousands of *maunds* (80 lbs.) being sent every season from some of the grazing districts to the more cultivated parts, especially to the western provinces. It is generally conveyed in dubbers, or large bottles made of green hide.

In Brazil there are four native modes of making butter. The first is, putting the milk in a common bowl, and beating it with a spoon as you would an egg. The second, pouring the milk into a bottle, and shaking it until the butter appears, when it is extricated by breaking off the top of the bottle, for bottles are valueless in that part of South America, on account of the number imported with fruits and liquids. The third, where the dairy is more extensive, is performed by filling a hide with the milk, which is lustily shaken by an athletic native at each end. The fourth by dragging the hide on the ground, after a galloping horse, until it is supposed that the butter is formed. The milk is never strained, and the butter never washed. The greater part of the butter used in the cities is imported from Ireland, France, or Germany, and this, notwithstanding that thousands of cows graze on the vast pampas in South America.

In Chili the butter is packed in sheepskins with the wool out, and would be very good in spite of appearances, were it not so much salted. The operation of churning is performed by a donkey; the cream is put into large gourds or dry skins, placed on its back, and then the animal is kept trotting round the yard till the butter is churned.

In Morocco the butter is churned by women in a bag of goat-skin, which, when nearly filled with milk, is closed by tying the mouth tightly. The bag is then rolled about and kneaded till butter is formed.

The principal countries in Europe which export butter are the following, with the shipments made in 1873:

	Cwts.
Holland	332,691
Sweden	69,815
Denmark	227,308

In France the butter exported in 1874 was 726,825 cwt., valued at more than £3,400,000. Canada exports annually about 140,000 cwts., and the United States a large quantity, of which we have no specific details.

The receipts of foreign butter in the United Kingdom in 1874 were larger than usual, and derived from the following countries:

	Cwts.
Sweden	23,292
Denmark	226,053
Germany	135,027
Holland	351,605
Belgium	76,723
France	713,251
United States of America	36,307
British North America	50,282
Other countries	7,268
	1,619,808

CHEESE, as is well known, consists principally of the casein of milk. The process of making it need not be described in detail here. Suffice it to say that the milk, at a temperature of about 120° Fah., is curdled by rennet, when the curd and a certain proportion of the fat separates, is strained from the whey, in which the milk-sugar still remains, and reduced by pressure to the more or less solid condition in which cheese is used. The chief differences in the various sorts of cheese are produced by the amount of cream or fatty matter left in the milk.

There is of course far greater variety in cheese than in butter, as cheeses vary from the pure cream cheeses made of cream only, which must be eaten fresh and will not keep, down to the skim-milk cheeses, which contain only a small proportion of fatty matter at all. Cheeses from skim-milk vary according to the number of skimmings; such are Dutch, Leyden, and Suffolk cheeses. Besides the differences caused by varying amounts of cream, are those produced by special methods of manufacture, by mixing cow's and goat's milk, etc.

The quality of cheese depends chiefly upon the milk of which it is made; the best containing a considerable portion of the constituents of butter. The Stilton cheese of England and the Brie cheese of France have a world-wide reputation, and are made from fresh sweet milk mixed with cream skimmed from milk of the preceding evening. The Cheshire, double Gloucester, Cheddar, Wiltshire, and Dunlop cheeses of Great Britain, are made of sweet unskimmed milk; as are also the best Dutch and American. Skim milk yields nearly as much cheese as sweet milk, as it contains all the casein. The ordinary Dutch, the Leyden, and the hard cheese of Essex and Sussex counties, are made of milk thrice skimmed.

Although the usual Dutch cheese is made of the skimmed milk, there are finer kinds made for home consumption, with a portion of the cream left in the milk. The most prized is the "schapekase," or ewe's milk cheese, which sells at a higher price. Gouda

GROUP OF HEREFORD CATTLE.

cheese, the best sort made in Holland, owes its peculiar pungency to the muriatic acid used instead of rennet for curdling the milk. Swiss cheese is usually made from skim-milk, and flavoured with herbs. There are, of course, richer cheeses, such as the Neufchâtel and Bondon, of the nature of cream cheese. In Westphalia cheese the curd is allowed to become slightly soured before it is compressed. The Italian cheese which is prepared for exportation is kept in brine, and is consequently excessively salt. It is only intended as a condiment for macaroni.

The poorer the cheese is, the longer it will keep, but every variety, if well cleared from whey and sufficiently salted, may be preserved for years.

Parmesan cheese owes its rich flavour to the fine sweet herbage of the meadows along the Po, where the cows are pastured. Dutch and Swiss cheeses contain, according to chemical analysis, from twenty to forty per cent. of nitrogenised matter, considered the most nutritive constituent of food. The best cheese is from twenty-five to a hundred per cent. more nutritious than bread or meat, which contains only about two per cent. of nitrogen.

The quantities of carbon and nitrogen in one pound of moderately good cheese are (according to Dr. Smith "On Foods") 2660 grains of the former and 315 grains of the latter, showing how rich this substance is in nitrogen.

To delicate stomachs cheese is objectionable on account of its slow and difficult digestion; but to individuals of great physical strength, it is a healthful and agreeable article of consumption. In combustible or heat-giving qualities, cheese is only exceeded by oil, butter, and like unctuous substances.

Cheese as an animal food may with advantage be substituted for butcher's meat at the current prices. There are good and substantial reasons for regarding cheese as a wholesome and valuable food, and it is worthy of even a more liberal consumption than it now receives. English people probably consume more cheese than any other nation on the globe, namely, in the proportion of about ten pounds yearly to each inhabitant. In the United States the consumption is about half that quantity. Besides being, when properly used, a wholesome and nutritious diet, and richer in nutritious value than butcher's meat or any other animal food, its peculiar ability to enhance the value and improve the healthfulness of the food with which it is consumed; the aid it renders in digestion, its readiness for use at all times without loss or trouble in cooking; its convenient form for handling and transporting; the ease and certainty with which it may be preserved for many months without loss or injury that occurs to other food from an excess of salt:—all these commend it to the favour of the public. In the army and navy, especially, it would be not only a luxury to soldiers and sailors, but a cheap, healthful, and substantial substitute for the continued use of salt meat.

The imports of foreign made cheese have been largely on the increase year by year, and have now reached about 81,300 tons while the home production is probably 100,000 tons.

The following figures give the imports at decennial periods, showing that our imports nearly double every ten years:

	Cwts.
1855	384,192
1865	853,277
1875	1,626,413

The countries from which we receive our supplies, are shown in the return for 1874:

	Cwts.
Holland	398,888
France	5,487
Germany	4,383
Belgium	1,625
Sweden	3,132
United States of America	849,933
British North America	221,043
Other countries	774
	1,485,265

K

The declared value of this quantity was close upon £4,500,000. Very little of this cheese was re-exported, nearly all being used in this country.

In Paris the consumption of cheese of all kinds, fresh and dry, in 1860 exceeded 8,000,000 pounds, which was an average of about seven or eight pounds per head of the population.

The principal cheese-producing countries are the United States and Canada, Holland, Switzerland, and Bavaria.

In the United States the production of cheese in 1850 was but 1,000,000 cwts., now it is more than double that amount; indeed the combined production of Canada and the United States in 1875 exceeded 3,000,000 cwts. It takes there a little over a gallon of milk to make a pound of cheese.

The exports of cheese from the United States in 1873 were nearly 1,000,000 cwts., and from Canada 174,000 cwts.; Holland exports from 500,000 to 600,000 cwts. annually; Switzerland shipped 392,153 cwts. in 1873.

The exports of cheese from France in 1874 were 182,353 cwt., valued at over £550,000, but all this was not French cheese.

Bavaria holds a position of importance for its production of cheese and butter. The cheese made is of a similar character to that of Switzerland, and is generally sold as such in Austria, France, and other countries. The production of cheese and butter from each cow on the dairy farms of the Allgars, the pastoral district, is computed at about 184 pounds, and the total production of cheese at 11,029 tons, and of butter 2,386 tons.

RENNET.—In cheese-making, the milk may be coagulated or curdled by the application of any sort of acid, but the substance which is most commonly used is the maws or stomachs of young calves, prepared for the purpose. These are generally denominated "rennets," but they are also often provincially called "vells," and in Scotland, "yearnings." In France the rennet is known as *présure*, or *caillette de veau*. Some people save the entire paunch, whereas it is only the fourth or true digesting stomach of the young calf that properly makes a rennet. The calf must be perfectly healthy, must have suckled the cow for at least four or five days, and to within a short time of killing. If it has been without food for any length of time, the stomach becomes inflamed, and especially so if the calf has been driven or carried a distance, and then it is of no value for rennet. The stomach should be taken out and well cleaned at once after the calf is killed, and, as soon as cold, is to be salted and left to dry on a dish for a day or two, then stretched on a hoop or crooked stick, and hung up to dry in a place where the temperature is moderately warm. The Bavarian method is to blow up the rennet like a bladder, and tie one end to keep out air, first putting on it a little salt at the place where tied. The skins being thus made very thin, will dry rapidly and keep well. Sometimes they are suspended in paper bags.

This prepared stomach, or rennet, when steeped in water, produces a decoction which possesses the power of thickening milk—decomposing it, and separating the casein from the liquid or whey. The most convenient way to prepare the rennet for use is to place the stomach in a stoneware jar with two handfuls of salt; pour about three quarts of cold water over it, and allow the whole to stand for five days; then strain and put it into bottles, or the rennet may be soaked over night in warm water, and next morning the infusion is poured into the milk. In from fifteen to sixty minutes the milk becomes coagulated, the casein separating in a thick mass. The rennet possesses the chemical property of producing lactic acid, by acting upon the sugar in the milk. The acid unites with the soda in the milk, which holds the casein in solution, when the casein, which is insoluble, separates, forming the curd. If we take an ounce of this membrane and wash and dry it thoroughly, and then put it into eighteen hundred ounces of milk, heated to 120° Fah., we shall find that in a short time coagulation of the milk is complete. If we remove the membrane from the curd, again wash, dry, and weigh it, it will be found that it has lost only about one seventeenth part of its weight. It is thus proved that one part of the active matter of the stomach has coagulated about *thirty thousand* parts of the milk. Hence we obtain some idea of the small quantity of rennet required to influence a large quantity of milk, and it would seem that comparatively little of the article would be required in the largest cheese factories. But such is not the case. The traffic in calves' rennets is immense in all cheese-producing countries, and the supply in North America from the millions of calves which are slaughtered is found wholly inadequate to meet the demand. The home supply falls short of furnishing to their cheese factories enough rennets by several millions annually; and consequently they are largely imported from Europe, but from all sources a sufficient supply of good, sound, healthy rennets is scarcely obtained. In France a good deal of rennet is procured from Switzerland.

An animal product has of late years become utilised under the name of Pepsine, which is the dried mucous membrane of calves' or sheep's rennets, the stomach of pigs and other vertebrate animals; it contains the active principle of the gastric juice, and is medicinally recommended to assist weak digestion.

CHAPTER IV.

OTHER ECONOMIC PRODUCTS OF CATTLE.

Having dealt with the food products of cattle, we come now to consider the various other substances of importance which they yield that are extensively utilised for Manufactures. This chapter, therefore, treats of Horns and hoofs and their commercial applications in the comb manufacture, and for horn buttons; we then pass on to Bones and their multifarious uses, the trade in Tallow and other animal oils and fats, and their application to the manufacture of soap, candles, and other purposes; thence we are led to a survey of the enormous commerce in Hides and the various stages of the Leather manufacture generally, and the tanning substances used, while parchment and vellum, bookbinding, and the miscellaneous applications of leather, are incidentally noticed. Finally the uses of offal, gut, and bladder, blood, cow-hair, and the glue manufacture are touched upon.

THE raw materials for Manufactures derived from cattle, which we import (exclusive of the large home production), consist of about 1,300,000 cwts. of hides, either in the dried or salted state; 6,000 tons of horns and hoofs; 74,000 cwts. of cow hair; 92,000 tons of bones; and 1,250,000 cwts. of tallow (some of which is sheep's tallow). Of the raw hides (exclusive of leather prepared in any way), the ratio in which the different countries contribute is, in round numbers, as follows: the South American States, 408,000 cwts.; India and the Straits Settlements, 320,000 cwts.; Europe, 267,000 cwts.; the United States, 126,000 cwts.; South Africa, 82,000 cwts.; and Australia, 22,000 cwts.

HORNS AND HOOFS.—The horns of animals, wild and domestic, may seem of but secondary importance at a first glance, and yet the trade in them rises to a very respectable figure in the statistical returns.

Our annual imports of horns, horn tips, and hoofs, average now 6000 tons, valued at about £173,000, besides the supply from our domestic cattle. This is double the amount of our imports a quarter of a century ago.

The study of the composition, formation, and growth of horn is an interesting one and well deserving careful investigation, in view of the manufacturing purposes to which this material may be applied.*

In common parlance any hard body projecting from the head, terminating in a free unopposed point, and serviceable as a weapon, is called a "horn." But the composition of these differs materially. Professor Owen well observes:

"The weapons to which the term 'horn' is properly or technically applied, consist of very different substances, and belong to two organic systems, as distinct from each other as both are from the teeth. Thus the horns of deer consist of bone, and are processes of the frontal bone; those of the giraffe are independent bones or 'epiphyses' covered by hairy skin; those of oxen, sheep, and antelopes, are 'apophyses' of the frontal bone, covered by the corium, and by a sheath of true horny material; those of the prong-horned antelope consist at their basis of bony processes covered by hairy skin, and are covered by horny sheaths in the rest of their extent. They thus combine the character of those of the giraffe and ordinary antelope, together with the expanded and branched form of the antlers of deer. Only the horns of the rhinoceros are composed wholly of horny matter, and this is disposed in longitudinal fibres; so that the horn seems rather to consist of coarse bristles compactly matted together in the form of a more or less elongated sub-compressed cone."

It is commonly believed that the horns of the ox acquire an additional ring every year after the third, but the addition of annuli is far from being regular in other species. Many rings are

* There is a very fine and extensive collection of horns and heads of all the principal ruminants shown on the walls of the Bethnal Green Museum.

gained in one year's growth of the ram's horns, and in those of some antelopes.

The length of the horn forms a distinguishing characteristic in some breeds of cattle; but whatever improvements may have been effected in the form and character of the carcase by the modification of food and habits, it does not appear that we have been able to superinduce any improvement or alteration in the size or texture of the horns. Indeed the horns of wild animals would seem to be more prominent than in the domesticated races. Some African tribes, such as the Makololo, are in the habit of shaving off a little from one side of the horns of their cattle when still growing, in order to make them curve in that direction and assume fantastic shapes. The stranger the curvature, the more handsome the ox is considered to be, and the longer this ornament of the cattle pen is spared to beautify the herd. This is a very ancient custom in Africa, for the tributary tribes of Ethiopia are seen on some of the most ancient Egyptian monuments, bringing contorted-horned cattle into Egypt.

The rights and privileges of the "horn-workers" and "horn-pressers" in former times occupied the prominent attention of the Legislature. But there is no fear in the present day "of the trade being ruined, and the business lost to the nation," as was the cry when the statutes 6 Edward IV. c. 1, and 7 James I. c. 14 were passed, forbidding the sale of horns to foreigners, and prohibiting the export of our wrought horns.

The invention of horn lanterns has been by some ascribed to King Alfred, who is said to have first used them to preserve his candle time-measurers from the wind. The Romans preferred lantern lights of the horns of the wild ox to others. They also used thin skins and closely shaved hides for lantern leaves, which, Martial says, very much resembled those of horn. A lantern in the last century was an indispensable family article; there was no going into the yard or out of the door on dark nights without one. A piece of horn was sometimes placed over the title of mediæval MSS. to preserve the letters from injury, while the transparent material allowed them to be read. The child's horn-book of later times had its leaves of alphabet and spelling covered entirely with thin sheets of this material.*

Although the principal manufacturing applications of horn are for combs, umbrella tops, and knife handles, yet there are other uses as extensive and varied as the descriptions of horn which come into the market, or bristle on the head of the animals characterised by these frontal appendages. Ox, buffalo, and deer horns, are those mostly worked up, but the horns of the rhinoceros, ram, goat, and some other animals, are also employed to a limited extent for different purposes. Rams' horns are sometimes made into snuff holders, or mulls, for the Scotch. One of these will be found in Case **168**. Their characteristic appearance is well known; we give on the opposite page a representation of the fine horns of the white-breasted Argali of Thibet, which bear a close resemblance to them. We shall here speak chiefly of the horns of cattle, leaving the others for description when we come to treat of the animals producing them.

The horns of the ox and buffalo are never shed, they are deposited in layers or bony cores, so that their general form is conical. Horns of various kinds form an extensive article of export from India; in 1872, 97,000 cwts., valued at more than £65,000, were shipped from Indian ports; and in 1873 the quantity was even larger, the value being £94,694.

The immense horns of the African ox, or Cape buffalo, and of the Java buffalo, and the Arnee buffalo of India, are the most valuable, and the extent of the trade in this class of horns may be estimated from the fact that about 2500 tons are annually received from British India, and 350 tons from the Straits Settlements, exclusive of those from Java and the other islands of the Eastern Archipelago, and these would represent a slaughter of 2,000,000

* Specimens of these may be seen in the Educational Court, South Kensington Museum.

cattle annually. The horns of the tame buffalo are much smaller than those of the wild animal.

From 800 to 900 tons of horns are received from the United States, and large imports from South America, and Australia. About one fifth of the supply of ox and buffalo horns is used up for comb making, and some for knife and cutlass handles, while a small portion is made into shoe lifts, scoops, cattle drenches, drinking cups, &c. The solid tips and the hoofs of cattle,

ARGALI (OVIS POLI).

which are composed of the same material as horn, are pressed into buttons.*

About 400 tons of horns are received annually in England from the River Plate; 1000 horns are usually reckoned as a measurement ton in shipping, but they are frequently freighted by weight; it will take nearly 2000 to weigh a ton. Those from Spain of a light yellowish colour serve to imitate tortoise-shell; the imitation is effected by solutions of gold, silver, and lead.

Mr. Hadfield in his "Travels in Brazil," tells us that at Rosario and Santa Fé on the Parana, streets and roads are repaired with

* Cases **168** and **169** are devoted to ox-horn and buffalo-horn applications.

heads and horns of cows and horses. In the Pampas the skull of a horse or cow serves for a stool, a chair, or a pillow, as the case may be. In a certain district in the suburbs of Lassa, the capital of Thibet, the houses are built entirely with the horns of cattle and sheep. These odd edifices are of extreme solidity, and present a rather agreeable appearance to the eye; the horns of the cattle being smooth and white, and those of the sheep black and rough. These strange materials admit of a wonderful diversity of combinations, and form on the walls an infinite variety of designs. The interstices between the horns are filled with mortar. Great pyramids of horns and bones have been formed on some of the prairies of North America by the hunters.

At one of the branches of the Upper Missouri there is such a pyramid, 18 feet high by 15 feet in diameter, made of elk horns, every hunter who passes making a practice of contributing his quota to the stock by way of good luck.

MANUFACTURING APPLICATIONS OF HORN.—While many of the former uses of horns for glazing purposes, for drinking cups, for horn-books, and for the bugle of the bold forester, have passed away, other and more elegant and varied applications have been found for this plastic and durable substance. Extensive as is the present use of horns, we believe that many further manufacturing purposes may be found for them, and that they will become even still more important in a commercial point of view.

They receive a great variety of applications at the present day, owing to their toughness and elasticity, as well as their remarkable property of softening under heat, of welding, and of being moulded into various forms under pressure.

To apply horns to manufactures they are treated as follows:—They are first thrown into water, and slight putrefaction commences, by which ammonia is produced, when the horn begins to soften. To carry this action further, the horns are transferred into a slight acid bath composed of nitric and acetic acids, with a small quantity of various salts. When the horns are sufficiently softened, which requires about two weeks, they are cleaned and split into two parts by means of a circular saw, and these are introduced between heated plates, and the whole subjected to an intense pressure of several tons to the square inch. The plates may be moulds, and thus the horn can be compressed into any required shape. A great improvement has of late years been effected in this branch of manufacture, which consists in dyeing the horn various colours. To accomplish this, the horn is first dipped in a bath containing a weak solution of salts of lead or mercury, and when the horns have been thus impregnated with metallic salts, a solution of hydro-sulphate of ammonia is rubbed on them, when a black or brown dye is produced.

Another method consists in mordanting the horn with a salt of iron, and dipping it in a solution of logwood. Very beautiful white fancy articles have been produced from horn, by dipping it first into a salt of lead, and then into hydrochloric acid, when white chloride of lead is fixed in the interstices of the horn, which then simply requires polishing.

COMB MANUFACTURE.—The most important use to which horn is put is for the manufacture of combs, and the annual value of horn combs made in this country is estimated at 400,000*l.*

The comb manufacture is pre-eminently conducive to national wealth, because therein the value of the raw product is greatly multiplied. The skilled labour placed on tortoise-shell increases it in value about 40 per cent., while horn (the generally used product) so favoured, advances 200 per cent. This latter, rough, unattractive substance, is split and heated, bent and planed, triturated and polished, pressed and carved and fretted, till at length it is sent forth into polite society, reduced to the most fairy-like proportions, elegant in its surroundings, having a highly polished exterior, a beautiful set of teeth, a graceful bend, and an elastic spring. Case **169** shows comb manufacture.

It is the laminatory character of horn that prevents the

economical use of mechanical aid to any large extent. The difficulties thence arising and hitherto insurmountable are an erratic and diversely running grain, the raising up of the fibres after every use of the file, saw, plane, or other cutting instrument, and therefore the necessity for constant removal of *débris* and dust from the product-face, and of continual polishing and gauging. This latter care is needed, because the original start has to be made with a thickness of horn much stouter than is needed for the perfect comb, to allow for the waste of manufacture. An additional difficulty is the requirement of heat in all the processes, and that continually. These and other causes have ever prevented the use of what may be termed perfect mechanical appliances in this trade industry, in order to elegant, complete, and rapid production.

Let us first enter the press house. All around on our right and our left lie heaps of horns, with the tips cut off, or divided lengthwise; while the ammoniacal smell of burnt horn affects the eyes, palate, and nostrils. On one side of this shed, or outhouse, is an ordinary furnace, a sort of Tubal-Cain improvisation; and close by, in front, is a huge hammer, or kind of movable anvil, working between upright iron guides, the hammer or anvil raisable by a pulley. The process goes on thus. The workman in front of the furnace takes one of the tipless horns, (after it has been rendered pliable by heat) and with a common strong ripping knife splits open the horn lengthwise in the direction of the varying grain—in other words, he merely divides the horn by the grain throughout. For to cut across the grain would be objectionable. The split-up horns are then again warmed (in hot water and by fire), are opened out pretty flat, laid between cold iron plates, and pressed quite level by aid of the before-mentioned hammer, a few iron wedges, and an oblong iron-bound space, sunk in the furnace floor, in which plates and horn are placed. The above plan is adopted in the case of "non-stained" goods. When the goods are to be stained afterwards (in imitation of tortoiseshell, it may be), the heated, ripped-up, and opened out horn is placed between *hot steel* plates, and more highly pressed, so as to reduce the horn-plates in thickness and to destroy the grain of the material. Then by the aid of other processes the horn will take the staining requisite in the subsequent operations.

The machine room may be called the laboratory of the comb works. Blazing fires, revolving lathes, choking dust, and horny abominations and smells of all kinds, greet you on entrance. Here the horn may be seen in all shapes and progress of development, receiving its direction, contour, polish, &c. The cutting apparatus works like a simple copying machine. Place the horn plate on the bench beneath, put over the plate a cutter of the shape, size, and outline of any comb you may subsequently require, strike down the press, and the piece is stamped out imme-

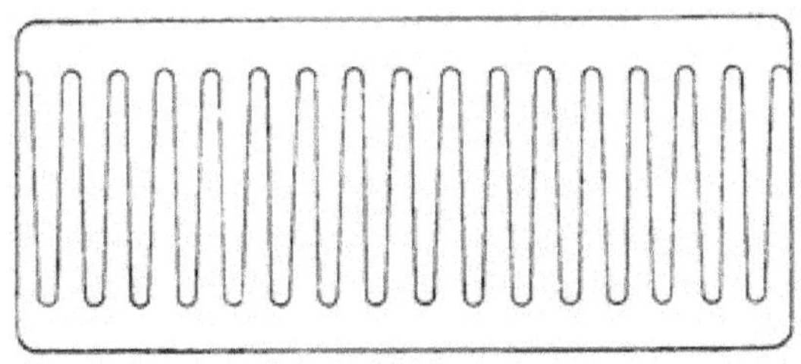

PARTED COMBS.

diately. Many pieces may, of course, be struck out by one die, and at one operation, the comb-plate being as economically used as possible. More pressing and straightening succeed, then grinding, ready for the "teeth." The mode of operation here depends on the kind of product you are manipulating. For a lady's back or side-comb, the "parting-engine" is put in requisition. This is a clever little contrivance, that cuts the teeth as it draws the horn-plate through the machine, working by a top handle also, like a copying machine. Each forward or backward motion of the handle brings down a tooth-cutter, and by means of a cogged wheel shifts on the bed on which the plate lies one tooth-distance further till all the teeth are cut. Various sized cutters may be used at one machine.

The last tooth at each end of the comb or combs is separated by hand, and then you have two perfect combs, the just-cut teeth fitting into and drawing away from each other, as the fingers of each hand, if they be placed the one between the other, for purposes of illustration. The teeth of horse combs and of those finer ones for the dressing-table, are cut by the circular saw. Suppose one (or more) very fine-toothed saw to be fixed on a rapidly-revolving shaft (lathe fashion) having a frame in front to hold the horn-plate or plates, to be toothed (for several in thickness may be done simultaneously, one lot in front of each saw fixed on the one shaft). This frame is centred or pivoted, so that it can be pressed close to, or be moved further from, the saw-edges; and has also a lateral motion acted on by a ratchet wheel.

Now take, say, a dressing comb, put on the ratchet wheel that will produce the number, or width and size of teeth you need (for all such wheels are numbered at so many teeth to the inch, and are made to suit the various sizes and shaped products), turn the handle, press down the horn pieces against the revolving saw, and (the pressure being regulated by the mechanism) the teeth are just cut as you want them—in depth, size, &c.—each backward motion of the frame from the saw sending the frame sideways just the distance needed to determine the width of the teeth;—thus this repeated action produces perfect teeth.

When the back of the comb is half straight and half curved, or in any other similar form out of the straight line, and the depth of the teeth has therefore to vary in accordance, the pressure of the frame (which holds the horn in process of "toothing") is increased or decreased against the saw, and so the cut is made deeper or less deep by causing the frame in its lateral progress to be assisted (in its proximity to the edge of the circular saw) by a projecting arm that is raised or depressed by its passage over a curved block or comb-back of the shape of the one in manipulation.

The fretwork in the back combs is all done by hand, the patterns being marked on the products and cut out by a very fine saw, a steady hand, and keen eye. The grooving and indentations on the comb-back are produced by the revolution of edged, grooved, serrated, and feathered wheels, against which the product is pressed to any required depth, exactly as the glass-cutter deals with his product. Thus also are the comb-backs of our dressing products channelled, grooved, roached, and otherwise adorned, the warmed and plastic horn being most obedient to every "good word and work." There seems scarcely an end to the rasping, planing, smoothing, and polishing, till, in the case of "stained" goods, the products are placed in a solution of weak aquafortis, and dotted with a red paint-like composition, to be subsequently chemicalised and washed off, when the stains will remain *à la* tortoiseshell !*

HORN BUTTONS.—The manufacture of horn buttons is one that is interesting from its antiquity, and from the modern improvements that have been introduced into it. Long before metal buttons were made for general wear, the horn button shared with the bone button the patronage of the poor. At present the horn button is used principally for shooting coats and vests, and for shoes and boots. The horn button is not, however, made from horn, as the name indicates, but from the hoofs of horned cattle. The hoofs of each animal will weigh about two pounds, and more than 80 tons are used up yearly by one factory in Aberdeen. The hoofs of horses are not suitable for the purpose. When the hoofs arrive, they are thrown into a large caldron, and boiled until they are soft. They are then cut into halves, and the sections transferred to the work-shop. Here the "blanks" are pierced or punched out by young women seated at hand-presses. The blanks, which are of a whitish colour, are then placed in vats in a strong dye, either of black, red, or green, the only colours which the hoof will take, where they remain till they are thoroughly

* Simmonds's "Technologist," vol. 5, p. 475.

dyed. Black is the most common colour used. The next operation is to fix the shank, which is done while the blank is soft and hot. This is a rapid process, and, like most of the other operations, is performed by children. The horn button, after being shanked, remains but a plain piece of rounded hoof, not even flattened or smoothed on its surface. The next operation is to place it in a mould, having an orifice for the shank to fit in. This mould merely contains the maker's name or trade mark, and is for the under surface of the button. The mould contains a dozen repetitions of the same pattern. When the buttons are ranged in this receptacle they are heated over an oven, till they are almost as soft as wax, when an upper mould, containing the pattern which is to be impressed upon them, and which fits closely upon the other, is placed over it. The two are then subjected to the press, and the buttons are taken out round and complete, with the exception of an occasional roughness round the edge, resulting from the overflow of the molten substance. This is afterwards pared off. The buttons are then fixed by their shanks upon a plate of metal, and subjected to the operation of a brush or a series of brushes, moved by steam power, which gives them the last touch, and produces a beautiful polish. They are now ready for carding and packing.

The Paris makers of horn buttons are celebrated for employing many of the best die-sinkers of France. Buttons of large size enable the artist to display ability, and the very fine classical heads upon some of the buttons in alto-relievo, show a great amount of skill in the execution of the dies; on some, colours have been introduced in indented rings, which at a short distance look both neat and effective. In some French buttons the opacity of the horn (or more properly hoof) renders them unattractive at a short distance, and it is only upon close examination that their merits are discovered.*

* The stages of manufacture, and the various illustrations of horn combs and horn buttons, can be advantageously studied in the cases **168** and **169** in the Animal Products Collection at the Bethnal Green Museum.

Bones and their Uses.—The bones of almost all animals are now articles of commerce; both wild and domesticated ones are made to yield parts of their osseous skeletons for some useful purpose. Thus we now import the bones of the giraffe, elephant, horse, ox, buffalo, and whale. The trade in bones has reached a very large amount, our imports exceeding 100,000 tons, of the value of £700,000.

In 1850 we imported but 27,198 tons, but in 1875 the imports were 104,971 tons, and those collected at home will probably be about the same amount. The greater quantity of these bones are used for manure. Of the imports in 1875, only 7,754 tons were suited for manufacturing purposes.

The composition of bones has been examined by many eminent chemists, but the most complete researches are those by M. Fremy. He found no marked difference between the bones of man, the ox, calf, elephant, and whale, but in the bones of carnivorous animals and those of birds there is a slight increase in the amount of mineral matter.

The general composition of bones may be considered to be as follows:—

Blood vessels	1
Osseine (usually erroneously termed gelatine)	32
Fat	9
Water	8
Phosphate of lime	38
Phosphate of magnesia	2
Carbonate of lime	8
Various other salts	2
	100

The fresh or green bones obtained at home are more valued than the imported bones, because they yield more gelatine and fat in boiling. Fresh unsteamed shin bones being solid, are very heavy, but as they become old they lose much of their weight. Old steamed bones are on the contrary light, and float in water. Bone is an important agent in many manufactures, being used by potters, turners, cutlers, gluemakers, sugar refiners, assayers, and by farmers for manure.

About 2,000,000 shank bones of oxen are worked up every year in Sheffield, &c., for knife-handles and spoons, and they are made into tooth and nail brushes, combs, fans, bone flats for button moulds, and various miscellaneous articles.

The small article of bone buttons constitutes an enormous trade. Prodigious quantities are made at Birmingham and Sheffield, and sent away in ton loads annually. So cheap are they,

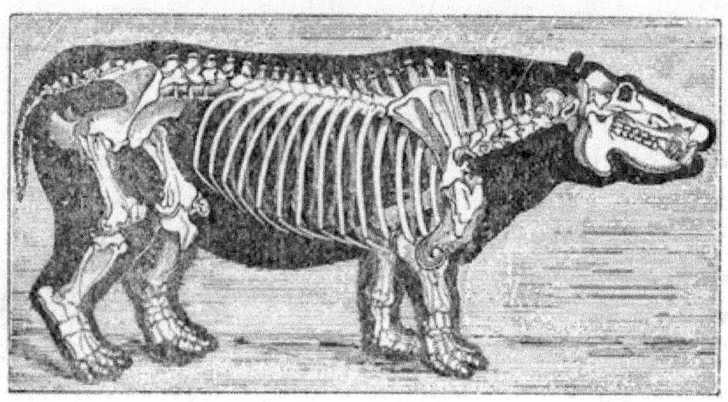

BONY SKELETON OF HIPPOPOTAMUS.

that they are sold at 4*d.* the gross wholesale, and hence can compete with any other material, even wood.

In the several stages of the useful applications of bones, we have first the shank and cut bones just alluded to for working up, next carbonised bones, burnt in closed air-tight retorts for about twelve hours, which are then ground between grooved plates to make animal charcoal in grain. This is employed as a filtering substance to clarify sugar in the process of refining. The portions of the carbonised bones which, in the process of grinding become too fine for filtering charcoal, are reduced to an impalpable powder and sold as bone-black to the blacking-maker. The grease or fat extracted from the bones before carbonising is used for making soap. Sulphate of ammonia, the liquor distilled out in the process of carbonising bones, afterwards saturated with sulphuric acid and evaporated, is used as smelling salts, and largely as a valuable manure.

Bone ash, when ground to moderately fine powder, is the material of which the cupels of the gold and silver assayers are made, being at the same time very infusible and sufficiently porous to discharge the litharge and other impurities, while the fine metal remains on its surface.*

THE TRADE IN TALLOW.—Of the different animal oils and fats, we shall only speak in this place of tallow, bone fat, and neats-foot oil, leaving horse-grease, lard, &c., to be dealt with in other sections.

Under the name of fats we designate the greasy matters, soft or solid, which fill in animals the cavities of the cellular tissues. It is found coating the intestines, round the kidneys, under the skin, between the muscles, near the base of the heart, and in the various cavities of the bones.

The composition of the fresh fat of animals, previous to its becoming rancid—that is before the formation in it of certain acids through the oxidising agency of the atmosphere, is oleomargarin and stearine, together with the membranous cellular tissues, in which these fatty substances are deposited. If we subject tallow to a considerable pressure we obtain a solid residue and a liquid oil, and even the most fluid oils contain a certain portion of solid fat.

The fat of various animals has not the same degree of solidification, fusion, or density; at different ages of the animal all these vary, becoming more elevated as the animal increases in age;—food, climate, work or action, modify these. The density, degrees of fusion, and richness in stearine of fats also varies

* The various economic applications of Bone are shown both in the Animal Products Collection, Case 170, and the Waste Products Collection of the Bethnal Green Museum.

according to the parts of the animal from which it is taken. The point of solidification of beef suet is about 98·5.

The tallow of commerce is the fat obtained by melting the suet of the ox or the sheep, and straining it so as to free it from membrane. When pure, it is white and nearly insipid, but as usually imported it has a yellow tinge and is classed according to the degree of its purity and consistence into candle and soap tallow. The tallow for making soap is considered very good if 13 cwt. of it will yield, with the other ingredients used, a ton weight of soap. The common kinds of soap are made from beef tallow and bone fat, and the better sorts from mutton and mixed tallows. The home production of tallow has been estimated as high as 120,000 tons, which added to the imports of 50,000 tons, makes a total of 170,000 tons of the raw material employed in the kingdom.

In 1840 the imports of tallow were 1,148,192 cwts., nearly all from Russia. In 1850 the imports from Russia were 854,144 cwts., and in 1875 a little more than 50,000 cwts. The United States, Australia, and South America are now taking the place of Russia in the supplies they furnish to our commerce, as will be seen from the following figures; but the Russian tallow still maintains a small superiority in price.

FROM	1860.	1870.	1875.
Russia . . . . . Cwts. .	1,082,663	242,541	50,517
Australia	12,005	489,751	270,498
South America	146,961	474,145	226,006
United States and other Countries .	188,479	316,861	420,375
	1,430,108	1,523,298	967,396

The sum paid annually for foreign tallow in this country ranges from £2,000,000 to £3,000,000, and very little is re-exported.

Tallow arrives at St. Petersburg from Siberia and the Ukraine from August up to the closing of the canals by ice. The quality depends upon its purity and its point of fusion. The point of fusion of the Ukraine tallow is 35° Réaumur. The initials P. Y. C. and Y. C. used in trade reports represent pale yellow candle, and yellow candle tallow. That imported from Russia is nearly all beef tallow; from South America and Australia a considerable quantity of mutton tallow is received, which is white, and fetches a rather higher price. New tallow will realize *1s.* or *2s.* a cwt. more than old. Case **158** contains tallows.

Town tallow is equal in value to Australian sheep tallow; melted stuff, or mixed kitchen fat and bone grease all have a value, although below that of the finer kinds of tallow. White bone fat is obtained by boiling or steaming fresh butchers' bones, and brown bone fat from street bones. The dregs or membranous refuse remaining after melting down tallow, is sold for feeding dogs, under the name of tallow greaves. Tallow oil is the oleine or liquid fat obtained in steaming and pressing tallow in woollen bags.

It was about the twelfth century that tallow torches came into use, and in the following century the tallow candle was generally employed, much in the same form and shape which it bears at the present day, but cotton being unknown, a flaxen wick was used. These candles in the time of the Romans were considered a great luxury, and only used by persons of high rank. In 1851 it was estimated that the average make of tallow candles, dips, and moulds, in this country, was 1,000 tons per week. The employment of gas in the present day, the cheap and abundant petroleum oil, and the large supply of vegetable oils, with the extensive production of stearine candles, have reduced the demand for ordinary tallow candles to a minimum.

The annual production of tallow in Russia about ten years ago was officially estimated at 160,000 tons, of which one half used to be exported, and the rest was locally consumed. Now the shipments are much smaller and the home consumption more considerable. There are in the empire upwards of 700 large tallow melters, who employ about 7,000 workmen. There are

600 candle factories employing 2,500 workpeople; 320 soap works, giving employment to 1,200 workmen, and thirteen steam works employing 2,100 people. The total value of the products worked up in Russia from tallow, exceeds £3,250,000. In Russia an ox in good condition will yield 200 to 250 lbs. of tallow. In Australia about 186 lbs. of tallow is obtained from each head of cattle, and 20 lbs. from each sheep.

At one time there was so little demand for cattle and sheep in Australia, that they were chiefly boiled down for their tallow. In 1848, 38,642 cattle, and 286,392 sheep were thus slaughtered in New South Wales for boiling down alone.

From 1844 to 1849, both inclusive, the number of sheep boiled down in New South Wales and Port Phillip (now Victoria) was 1,565,752, and of horned cattle 184,064, producing 440,186 cwt. of tallow.

In the six years ending 1865, an average of 37,829 head of cattle were annually melted down for tallow in Victoria, and in 1868 and 1869, 410,048 sheep were thus disposed of. The mode of doing so is the following:—The carcasses of sheep are chopped up in three or four pieces, and placed in large vats, eleven feet high, capable of holding 300 to 400 animals, and steam being applied, the fat swims on the top, and is drawn off into coolers, which ordinarily contain 500 gallons. The flesh is given to pigs or made into manure.

An animal oil which is much in request is neat's-foot oil, made by boiling down the feet of cattle, by tripe-dressers and others, who purchase the offal of the markets. Ten of these feet will yield about a quart of oil. This oil is retailed at about 5*s.* a gallon. Some quantity is shipped from the River Plate and the Falkland Islands, and a good deal is made in London, Paris, New York, and other large cities. As it remains liquid below 32°, and is in other respects a useful lubricant, it is much employed to oil machinery, church clocks, &c. This oil can be easily purified and freed from the deposit by placing it in the sun's rays with a few strips of lead in the vessel, when the pure oil will rise to the top.

SOAP.—Soap is the product of the action of caustic alkali upon neutral fats. Potash and soda form soluble soaps; of these potash forms soft soap, and soda hard soap. The latter is consequently most generally employed. The cleansing property of soap is usually considered to depend upon the amount of alkali which it contains; pure alkali would injure the hands and the fabric, but by combination with the fatty acids its action is rendered milder without destroying its property of combining with impurities, and especially with fatty matters. Soap is used for cleansing purposes, in washing, in bleaching cloth and woollen materials, for the preparation of lithographic tints, &c.

The animal substances principally employed are, imported beef and mutton tallow, horse-grease, lard, and fish-oils, also bone-grease, kitchen refuse, town tallow, and fat obtained from butchers. Lard is rarely used in Europe for making soap, being too dear; but it is extensively employed in the United States, where enormous quantities are converted into a solid fat and a fluid oil.

The greater the quantity of fat acids combined in the soap the higher is its value. It is in the power of the soap-maker to manufacture 300 parts of a good hard soap out of 100 parts of fat. When but a small quantity of water is contained, the soap becomes very hard, and much labour is lost in obtaining a lather. If on the other hand water is held in too large a proportion, there is a great loss of material. For the common yellow soaps and for soaps for sizing paper, resin is added to tallow in the proportion of 50 per cent.

The soaps made in England are distinguished by a good substance, a soft cut, and a composition appropriate to the special uses for which they are made—and for their cheapness. The best are well saponified without excess of alkali or salts. The materials employed are chiefly tallow for white soap, and bone grease and kitchen stuff for marbled soaps and Windsor soaps.

Since the abolition of the duty on soap, and the soap-makers' licence, it is not possible to obtain any precise returns of the quantity of soap made. In 1851 there were 317 licensed makers in Great Britain, but by centralisation and larger factories the number of makers has been greatly reduced. The quantity of soap manufactured in 1851 was 100,000 tons, of which 24,500 tons were exported and used by manufacturers, leaving 75,500 tons for domestic use, a proportion of a little over 8 lbs. to each person. The manufacture has since largely extended and the product is cheaper, owing to the greatly increased supplies of vegetable oils. The export of soap, which was then (1851) about 6,300 tons, was in 1875, 12,538 tons.*

THE TRADE IN HIDES.—The demands of the civilised world bring all descriptions of hides and skins to our markets, such, for instance, as the miscellaneous collections of skins and peltries of the Hudson's Bay Company. At the public sales in London and Liverpool, held fortnightly, immense numbers pass under the hammer of the auctioneer. Strange mixtures are sometimes met with in the different lots offered. For instance, the Commercial Society of Mozambique sold at Rotterdam, in June, 1876, 5,461 skins of the gnu (*Catoblepus Gnu*), 2,542 of the quagga (*Asinus Quagga*), 436 of the giraffe (*Camelopardalis Giraffa*), 96 boar skins, 14 lion and hyena skins, 41 deer, 391 buck skins, 2,168 of the blesbok (*Gazella albifrons*), and 3,071 of various other antelopes.

Skins vary in texture and substance as much as the character of the animals they cover. Mr. W. N. Evans, a scientific and practised tanner of Bristol, tells us that the sheep, though valuable for its wool, gives but a spongy, porous skin, only available for common work, such as wash leather and imitation chamois. The pig, with a skin of a closer texture, is yet of so inferior a description as to be principally confined to saddler's work,

* Case **159** contains spermaceti and Case **160** soaps.

for which it is well adapted. The horse is the possessor of a remarkably thin skin, almost transparent; the naturalist would do well to notice one peculiarity about it, which is, that whilst three-fourths of the hide make the best curried leather, the remaining portion from the hip bones, covering the rump, has under the true skin a layer of thicker substance, giving to it an inflexibility very foreign to the rest of the hide. No doubt this layer was placed there for some wise purpose; but what that purpose is we have yet to learn. The hide of the ox is the most important. The calf and younger cattle give us skins which are mainly used for the uppers of boots. The ox at maturity furnishes the stouter hide, which the tanner transforms into sole leather. What are the chemical constituents of a hide? A transverse section of fresh skin shows it to be a gelatinous mass, the substance of which is full of fibres interlacing each other in every direction. The microscope reveals to us the corium, or true skin, and the cuticle, or epidermis; between the two lies the rete mucosum. The cuticle in which the hair is imbedded is removed during the process with the hair, leaving the true skin to be made into leather.

The skin in its normal condition, as received from the butcher in the home market, possessing its original moisture and flexibility requires but little skill to prepare it for the process of tanning. Our home product of hides was valued in 1873 at £7,000,000 sterling; but the imported hides and skins of different kinds received from all parts of the world in the same year, raw, tanned or dressed, exceeded £10,000,000 in value.

The vast Pampas bordering on the River Plate, in South America, now furnish large quantities of the best hides. The River hides, as they are called, consist of the Paysandu ox hides, which range from 60 to 67 lbs. on the average. The Gualeguaychu ox hides 62 to 64 lbs.; the Salto ox 66 lbs., and the Concepcion ox 64 lbs. The Buenos Ayres hides average about 62 lbs.

The hides while quite fresh are put into brine pickle for twelve to twenty-four hours, when completely saturated they are removed to drain, then salted and placed in piles for ten to fifteen days, after which they are ready for shipment. They are placed in layers on board ship for passage to Europe, with bay-salt between each layer, and reach our shores in a moist condition. A heavy River Plate ox hide may measure 7ft. long, by 5ft. 9in. wide. A light one 4ft. 10in. by 4ft. 4in. A salted ox hide may weigh from 42 to 89 lbs. According to size one thousand ox hides, with the necessary salt, may be taken to weigh about 30 tons. The number of hides, salted and dried, shipped annually from Buenos Ayres is from 2,500,000 to 3,000,000, from Monte Video 1,500,000, and from Brazil about 1,500,000. The principal markets for these hides are first France, second the United States, third Belgium, and fourth England.

Havre and Marseilles are the principal ports of import in France: at the former about one million of hides are received annually; at Antwerp one million and a half are imported.

Of the one million and a half hides shipped annually from Brazil three-fourths are wet salted, and one quarter dried. The former are sent chiefly from San José do Norte, and the latter from Rio Grande do Sul. The latter province used to swarm with wild horses and cattle, and nearly 600,000 head of cattle are slaughtered there annually. The Brazilian hides weigh from 30 to 32 lbs. each, and as freight 70 are computed to weigh a ton. From Guatemala 627,000 hides were shipped in 1874, and from Texas at least 300,000, either dried or salted, are sent away.

The excellent material obtained from horned cattle on the steppes, on the immense pasture lands of Podolia, and on the broad plains of Central Russia acquires a high degree of solidity, flexibility and excellence from the perfect system of tanning employed. There are in Russia about 3,000 tanneries and leather manufactories, and the annual production in prepared hides and sheep skins amounts to close upon £8,000,000. The number of raw hides and skins obtained annually is estimated at 20,500,000, of which 19,000,000, are dressed. Deducting sheep skins, which are more than 50 per cent of the whole, the numbers and weights may be taken as:

Ox-hides . . .	2,200,000	averaging	40 lbs.
Cow-hides . .	1,300,000	,,	20 ,,
Calf-skins . . .	4,000,000	,,	3 ,,
Horse-hides . .	1,000,000	,,	12 ,,

About one-tenth of these are exported.

There are certain technicalities in the leather trade which require explanation, thus we have the "crop" or full hide, the "butt" or rounded hide, and the "offal" which consists of shoulders, cheeks, and faces, necks, bellies, and middles. A side of leather is a "bend."

In trimming leather a very little is taken off the butt end, then about a foot in depth or less according to the quality or thickness of hide is taken from the belly, then from the fore end of the bend to within about six inches of the horns a square piece called "shoulders" is taken, and from this a piece is generally cut called the "range," which is from above the shoulders to the horns. The rest is "cheeks and faces." Thus we have from a whole hide five different kinds of leather, which bear about the following relative prices :—

	Per lb.—*s.*	*d.*
Butt	2	0
Belly	1	0
Shoulders	1	4
Range		11
Cheeks and faces . .		9

The oblong portion between the two belly parts marked G G, is known as the "butt," and when split down the ridge, as shown by the dotted line down the centre, the two pieces are known as "bends;" the two pieces marked Y are "belly offal;" D is known as "cheeks and faces." The butt within the dotted line may extend in length from A to B or from A to C; if cut off between B and C that portion is called the "range;" or the whole from B to X may be cut in one piece and termed a "shoulder." Sometimes the range is cut off, and the rest would be called a shoulder, with "cheeks and faces" on: or, again, the range and shoulder may be all in one nearly square piece. The manner of cutting this part depends upon the spread and size of the hide. The sketch on the opposite page shows what is termed a close rounded butt. There is generally more belly taken off an English than from a South American hide, and for strap leather they round closer than for the other purposes. The belly taken off a River Plate hide is narrower than what is shown in the diagram; the most pains is taken in tanning the butt, as they always tan offal by the quick process. When leather is trimmed in this way the belly offal generally sells at one half the price of the butt or bend, the shoulder a few pence below the price of the butt, and cheeks and faces about half the price of the shoulder.

Crop leather weighs from 26 to 36 lbs., and shaved hides from 16 to 26 lbs., and sells at some fourpence per pound more than the crop. The rounded shaved leather, generally cut into sides, is dressed leather and sells always by the pound.

Calf skins vary in weight from 20 to 25 lbs. up to 100 to 120 lbs. They are sold either entire or rounded. The heavy ones are not so valuable as the light ones by threepence to sixpence per lb.

"Kips" is the name given to the ox hides while in the state of transition from the calf to the fully developed animal, this applies to English, St. Petersburg, and South American Kips. But there is another kind of Kip imported in large quantities from the East Indies, which are the skins of a small breed of oxen shipped from the three Presidency ports of Bombay, Madras, and Calcutta. A Patna Kip is the hide of a two year old steer or heifer, and is supposed to be anything in the shape of a hide weighing less than 16 lbs. As a general rule Kips come to market salted. When properly tanned and dressed, these hides make fine leather, ranking next to calf skin. They are shipped from Calcutta, both dry and dry salted, folded down the back, and also plastered over the flesh with some kind of whitewash. Kips range from $3\frac{1}{2}$ to 9 lbs. Of late years the greed of commerce has induced unscrupulous

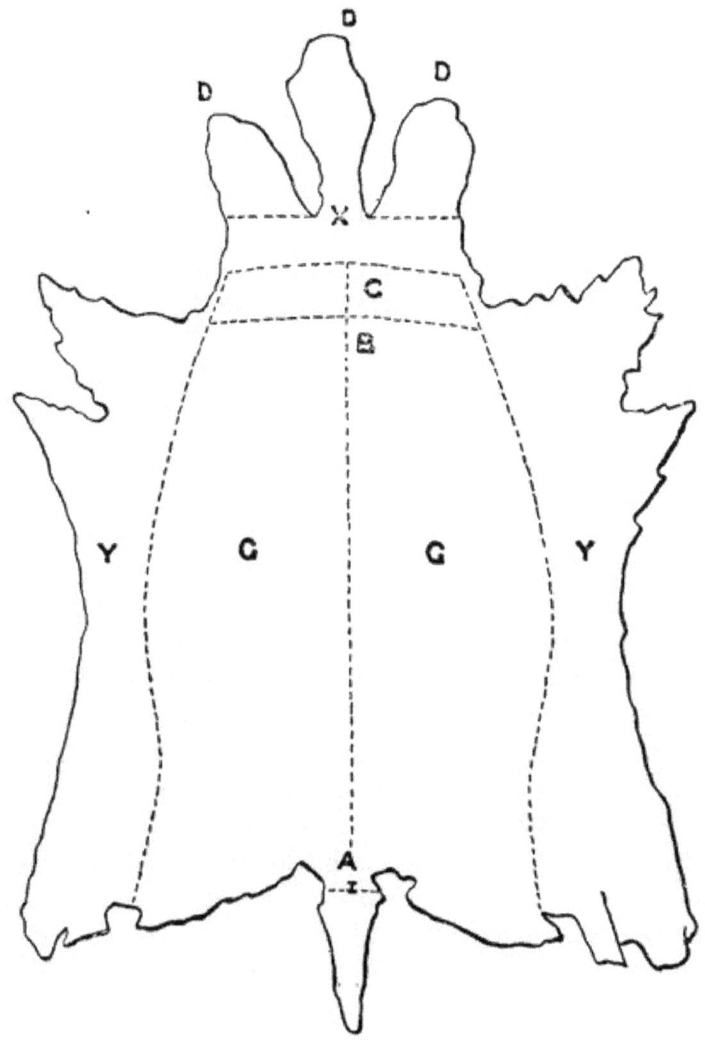

CROP, OR FULL HIDE.

dealers to daub the fleshy side of the Indian Kips with large quantities of plaster, as much as 3 lbs. being often added, and the average is $1\frac{1}{2}$ lbs. In the five years ending 1868 the annual value of the hides and skins shipped from India was £900,000, in the next five years it was £2,560,000, and the average of the three years ending in 1875 was £2,739,134.

The following shows the number of hides and skins shipped from India in 1873 and 1875:—

	1873.		1875.	
	Tanned.	Dry and Salted.	Tanned.	Dry and Salted.
Bengal	1,049,829	9,788,504	6,775,920	9,636,651
Bombay	2,040,735	1,015,826	1,019,201	270,712
British Burmah	3,352	367,840	3,600	274,029
Madras	8,119,222	271,071	6,467,764	148,087
Scinde	6,044	325,192	17,562	80,143
	11,219,182	11,768,433	14,284,047	10,409,622

The total number shipped in 1874 was 19,295,552. There is an export duty in India of 3 per cent. on tanned hides and skins; raw hides are shipped free.

The cows, bullocks, and buffaloes in the Punjab in 1875 numbered 6,570,212.

LEATHER MANUFACTURE.—Next to wool, the leather trade forms one of the largest British industries connected with animal products. By the census of 1871 there were in England and Wales alone more than 55,500 fellmongers, tanners, curriers, and workers and dealers in skins; 25,000 workers and dealers in saddlery, harness, portmanteaus, &c.; 21,000 glovers; and nearly 225,000 boot and shoe makers. The uses of leather are so numerous that it is quite impossible to estimate the home consumption. Even if it were calculated that one half of the population of Great Britain used only two pairs of boots or shoes each yearly, at 5*s.* a pair, this would give an annual outlay of £16,368,700. This sum might, however, safely be doubled, to say nothing of the gloves, saddlery, belting, and other multifarious uses of leather.

Enormous progress has been made in the leather trade. English leather of the best quality is now supplied to the principal markets of the Continent, and bids fair to hold its ground. While in 1828 the value of the leather of British manufacture exported, was under £164,000, in 1875 it had risen to £3,874,108.

The Animal Products Collection at Bethnal Green Museum is very rich in hides and leathers from all countries, obtained at the various International Exhibitions; there are numerous illustrations of the applications of leather for book-binding, saddlery and harness, wall hangings, shoes, gloves, and other purposes.

Tanning Substances.—For the use of the 800 Tanneries of the United Kingdom, and to work up more than 190,000 tons of hides and skins into leather, a large quantity of Tanning Materials is required. The principal supply is obtained from the bark of our indigenous oaks, and from the larch tree in Scotland. Mimosa, Cork, and a few other foreign barks are imported, and bark extracts from North America, to the value of £64,000. Gambier, Sumach, "Divi Divi," (a leguminous pod from Rio Hache), Myrobalans (a dried Indian fruit), and the acorn-cups of a species of oak of Southern Europe, passing in commerce under the name of "Valonia," are the other tanning substances chiefly used.

The proportion for producing one ton of leather would be, of Bark 6 tons, Sumach or "Valonia" 2 tons, Gambier and Cutch 1 ton. The following figures will give an idea of the aggregate consumption in 1874, and the current prices:—

		Quantity.	Price.
British Oak Bark in loads of 45 cwt.			£ £
(Hatched Bark)		200,000	17 to 19 per load.
Foreign Oak Bark	cwt.	207,168	6 to 7 ,, ton.
Larch Bark	loads	50,000	5 ,, load.
Mimosa Bark	cwt.	117,168	12 to 14 ,, ton.
Valonia	tons	26,336	15 to 20 ,, ton.
Myrobalans	cwt.	338,466	13 to 16 ,, ton.
Sumach	tons	16,514	12 to 19 ,, ton.
Gambier and Cutch	tons	16,441	25 to 35 ,, ton.

* In Cases **136** and **137** will be found a large and varied collection of the vegetable substances used in tanning.

Parchment and Vellum.—The origin of parchment is very old, the skins of animals prepared for writing on being generally used before the discovery of paper. Herodotus tells us the Greeks wrote on the skins of the goat and the sheep deprived of their hair and prepared; Josephus states that the copy of the Holy Scriptures sent to Ptolemy by the High Priest Eleazar was written on fine animal membrane. Pergame is generally said to be the locality where it was first invented, and parchment still bears this name in parts of the continent. Before the Revolution, the University of Paris had the sole right to sell parchment, and there was a street set apart for it. All made had to be sent to the Hall to be examined and stamped. The object was to secure the authenticity and date of acts and deeds, &c. Parchment is usually made of sheep skin but can also be made with those of goats and calves. The skin of the ram makes a better parchment than that of the ewe, but the best is from lamb skins. Parchment for drums, tambourines, &c., is usually made of calf skin or goat skin; wolf skin is said to be the best for this purpose.

Parchment is much used for commissions, warrants, writs, diplomas, title deeds, and estate plans, &c., which require to be preserved. It is also employed by miniature painters, for church missals, in bookbinding, &c.

Vellum is a finer, smoother, white kind of parchment, made of the skins of young calves which are larger, thicker, and more difficult to prepare than sheep and goat skins, hence the higher price. It is elastic, stronger and more smooth and compact than parchment. It is therefore selected for manuscript addresses, designs, and church books, &c. Before the discovery of oil painting, the choicest drawings were made on vellum.

The mode of preparation is first to take off the hair or wool, then to steep the skin in lime, and afterwards to stretch it very firmly on a wooden frame. When thus fixed, it is scraped with a blunt iron tool, and wetted and rubbed with chalk and pumice stone, till it is fit for use. It is curious to notice that from about the seventh to the tenth century parchment was beautiful, white, and good, but in later times a very inferior, dirty looking kind of parchment came into use, which has the appearance of being much older than the good. The reason for this is supposed to be that the writers in these later centuries used to prepare their own parchment, while at an earlier date it was a curious art, only possessed by the manufacturers.

The scarcity and expense of parchment, and the demand for the writings of the fathers and books of devotion, in the middle ages, frequently induced the monks to erase or wash out the writings of the classical authors, to make room for those of the fathers. In many cases, however, they did not obliterate entirely the ancient writing, and a careful examination of some of these "palimpsest" MSS., has led to the discovery of valuable works and fragments of the classical authors, among the rest one of the works of Cicero.

Case 135 contains specimens of parchment and vellum, and the tools used in preparing it, and there are some framed specimens of very old parchment writings. There is in the case a specimen of parchment made from kangaroo skin. The refuse chippings and cuttings of the manufacture are used for making size.

BOOKBINDING.—Great improvement has taken place in all descriptions of bookbinding within the last few years, more especially in the application of colour and ornament to the cheaper kinds. So great has been the extension of this trade, that the introduction of machinery for performing the operations of embossing and blind-tooling became necessary in order that the bookbinder might keep pace with the demands on his trade. The editions of most popular works are now so large, and have to be published at so low a price, that the cheapest style of binding has to be adopted. An edition of a thousand volumes can be put up in cloth covers, lettered and gilt, by some of the large binding firms in six hours.

Mr. Reed, reporting on the Bookbinding exhibits at the London Exhibition of 1862, observed that in bookbinding there may be said to be three grand divisions: first, those works of art which in all the magnificence of splendid covers, jewelled ornament, and carved ivory, and enamel, seem to rival in richness the bindings of the earliest printed books, which on account of their rarity have been treasured, especially abroad, in the most costly covers, weighted and enriched by reliefs in gold, silver, and other chased metals; secondly, work marked by a degree of simplicity bordering upon rigidity, combining an elegance and solidity in design and manipulation which furnish the highest style of work, in which the finest relics of antiquity, the choicest reprints of rare books, and the highest class of modern literature, are encased; and thirdly, there is the bright, gay, and begilded cloth and paper binding of our ephemeral literature, rapidly produced, and as rapidly destined to destruction.

In all these sections there is a marked advance; in the first, extravagance has to some extent given way to utility; in the second, the exactitude of the mere copyist has yielded to the demand for some degree of originality in design and greater care in the combination of colours; and in the third, cheapness of production, and the valuable initiative faculty have given to us instead of clumsy and inartistic covers, a degree of merit in cheap work which is one of the marvels of the present age.

Half-binding is that style in which only the back and corners are covered with leather, and the sides with paper or cloth. In the finishing or ornamenting of a bound book much taste may be displayed. The edges of the leaves are usually either sprinkled with colour, marbled, or gilt with leaf gold. The covers are sometimes coloured or sprinkled by the binder, and are impressed both on the sides and back with ornamental devices and inscriptions, by the application of heated stamps or dies, either with or without leaf gold; such impressed devices as are not gilt being distinguished by the name of blind-tooling.

When gold is used, the surface of the leather is prepared to receive it by the successive application of parchment-size, white of egg, and a little oil. In ordinary handwork, the patterns are produced by the separate application of a number of small dies, and engraved rollers for lines and long narrow patterns, but sometimes a number of dies are fitted together and applied simultaneously by means of a press. This process is called blocking. Various shades of colour are applied to the leathers used in bookbinding. We have bindings in Russia and Morocco leather, fancy calf, goat, parchment and vellum, and embossed leather. Among the other materials used for book-covers are tortoise-shell, carved ivory, velvet, mother-of-pearl shell, oak, and other carved or inlaid wood.

Case **144** contains specimens of bookbinding.

MISCELLANEOUS APPLICATIONS OF LEATHER.—The leather manufacture is one of our most ancient and important industries, and there are very many minor and miscellaneous applications of leather which add variety and interest to a collection. Even if modern illustrations of shoes and boots are not necessarily shown in the Museum, some of the curious varieties are, such as alligator, snake skin, porpoise skin, and kangaroo skin boots. Samples of leather tankards, bottles, sandals, caskets, and shields as exemplifications of former uses; sheaths and scabbards, braces, breeches, vests and buttons, as articles of dress. Fancy leather work, as frames, flowers &c. There are also a host of small articles in leather work, such as port-monnaies, pocket-books, purses, card cases, sample-cases, letter-cases, cigar-cases, writing-folios, tobacco-pouches, cloak straps, ladies' belts or leather girdles, which in many instances are pretty and attractive. Leather palms used by the sail-maker, buckets, belting and driving straps for machinery, buff leather, a strong oil leather prepared from the skin of the ox, buffalo, elk, &c., and used for polishing wheels, buffsticks or glaziers, knifeboards, razor strops, &c., and other uses.

The old adage that there is nothing like leather is certainly verified in the multifarious uses to which leather has been or is now put. We make coverings of it in articles of personal use, for a man may be clothed in leather garments from the head to the foot. In saddlery and harness its use is universal, and nothing can supplant it for durability. In articles for household and domestic use, we have leather hangings and coverings for furniture, buckets and bottles, cups and hose. Eleven frames of handsome tapestry leather are shown. For travelling we have portmanteaus, valises and hand bags, pocket books, purses and cigar cases. In Case **145** are examples of many of these. We write on leather and we cover our books with it, and it has even been used by photographers to take likenesses on. It is the packing and baling material in many countries from its cheapness and durability. Hammocks, boats, and even cannon, have been made of it, whilst the leather apron is the most durable and serviceable protection for many an artisan. Leather shields were and are still in use in many countries. It serves for the grip handle of swords, and for sheaths of knives. We use leather in balls for cricket and football, and we cover musical instruments with it, as well as telescopes and many philosophical instruments for protection. It is the most ancient, useful, and generally applied animal substance for an infinite variety of purposes. And moreover leather can be made of the skin or hide of almost every quadruped, and of many fishes, serpents, and reptiles. Human skin has even been tanned, but it is too thin for any serviceable use. In Cases **140, 141, 142** are illustrations of leather work, and Case **138** contains specimens of tools used in the working and preparation of leather.

There are eleven large framed photographs shown, taken at Messrs. Bevington and Sons' tannery, Neckinger Mills, Bermondsey, which illustrate and explain the whole process of tanning and leather preparation, while other frames from the same firm contain samples of various manufactured leathers, and of the stages of progress in preparing sheep-skin, goat and calf-skin, seal-skin, and horse hide.

USES OF OFFAL.—What shall be done with the fifth quarter of the animal, or the "offal" was a question that formerly used to be perpetually assailing Boards of Health, and other sanitary bodies who have the supervision of slaughter-houses, meat-markets, &c.

Now, however, the offal of cattle suited for food, the waste from dressing skins and preparing leather, and other animal refuse, all have their distinctive and remunerative uses.

The tanner makes use of the hide, while the bones and horns are put to various uses. The suet is melted down by the candle makers, leaving only the head, offal, and blood to be disposed of. We shall now proceed to trace out some of the purposes in which these are employed. The allowance for offal is very largely and variedly influenced by the breed of the animal, sex, age, and accidental circumstances. The following figures may however, be taken as the average medium weight of the offal in fat cattle brought to market.

In general, hide and horns, 4 to 7 stone of 8 lbs.—in rare cases, 8 to 9 stone; tallow, 3 to 10 stone—in rare cases, nearly to 20 stone; head and tongue, 2 to 3½ stone; kidneys, 2 to 4 lbs.; back collop, 2 to 4 lbs.; heart, 6 to 9 lbs.; liver, lungs, and windpipe, 1 to 2 stone; stomach and entrails, 10 to 14 stone; blood 3 to 4 stone. Or the proportions may be taken to be as follows:

	Carcase.	Offal.
Store oxen	59·3	38·9
Fat oxen	59·8	38·5
Fat heifers	55·6	41·3
Fat calves	63·1	33·5

Much of the offal is used as food, such as the head, tongue, feet, heart, liver, lungs, tripe, &c.

The head of the ox contains about 30 per cent. of meat, which is usually fat, and produces a nutritious soup. Calf's head is a more delicate and expensive dish. Ox tongues are usually salted or pickled, but are sometimes dried, and there is a considerable commerce in ox tongues imported from Russia, Australia, and River Plate.

In parts of Africa, according to Dr. Schweinfurth, before the tongue of any animal is eaten, the tip has to be cut off, for here they say, is the seat of all curses and evil wishes; and even the tongues of sheep and oxen are not served up until they have been subjected to this treatment.

A profitable business is carried on in London and other large towns in animal offal for food. In the metropolis there are no less than 100 tripe-dressers, many of whom sell many varieties of animal offal prepared for food. The liver and the lungs, or as they are vulgarly termed lights, of some animals are eaten. The intestines are also used as food in the preparation of sausages and black puddings, whilst the thicker and fatter parts are eaten as tripe.

The feet of animals consist of two chief chemical constituents of food, viz., oil and gelatine, and hence we have neat's foot oil (already alluded to) and calf's foot jelly. The oil has too strong a flavour to be used as food, and must be removed before the foot is eaten. This is effected by the application of heat after the free use of the knife, and as the foot is cooked by being boiled in water, it is necessary that the oil should be skimmed from the broth, that the latter may be fit for food.*

Neat's foot oil, trotter oil, tallow oil, and all the other animal oils of commerce, are shown in the Animal Products Collection and Waste Products Collection of the Bethnal Green Museum.

Gut and Bladder.—There is a larger and more important trade than is generally supposed in the preparation and application of the intestines of animals for various purposes. They serve for the preparation of goldbeater's skin, bladders, rackets, lashes or whipcords, sausage skins, cords for clock-makers, hat-

* Dr. E. Smith "On Foods."

makers, grinders and polishers, and strings for musical instruments. Thirty years ago the value of these in Paris was set down at £50,000 a year, and in London, from the much more extensive slaughter of animals, a far greater trade is carried on.

The membrane of the intestines or blind-gut, of which goldbeater's skin is made, is remarkable for its strength and delicate fibrous construction. The manufacture, important as it is, rests in the hands of a very few manufacturers. The principal London maker uses up the skins taken off the gut of 10,000 oxen weekly.

Great care is required in the manipulation and successive processes of drying, bleaching, and preparing, these skins being of the highest importance to the goldbeater, for the beating of gold and silver leaf. The gut or skin is stretched on frames, and to give them greater support two skins are doubled together, after which they are well cleansed and scraped to remove all superfluous fibrous matter, and being well rinsed and dried are coated with a varnish. In its finished state this skin sells for ten to twelve guineas the mould of 850 small leaves, such a mould enabling the workman to beat out that quantity of gold leaves at one time. Each leaf in the mould being double, there are 1,700 thicknesses of skin.*

These membranes are of three kinds, an outside covering of common parchment, a set of leaves made of very fine and smooth calf-skin vellum, and another set of goldbeater's skin.

Bullock's weasands or ox guts are cleaned and salted for exportation, to be used as covers to sausages and polonies. Besides the bladders obtained from our own slaughter-houses, we import a good many from the Continent. The imports of bladders in 1870, the latest year for which there is any official record, were 652,361, valued at nearly £3000.

Bullock's "bungs," as they are technically termed, are the

* This manufacture is shown in case 173.

largest, and more shapely than any other bladder. Bladders when brought to a clean and prepared state are specially useful to druggists, oilmen, colourmen, and other manufacturers as coverings for various kinds of vessels; they derive their value from their thinness, toughness, and impermeability to water. The bladders of the ox and other animals, when deprived of bits of loose membrane and other impurities, are washed in a weak solution of chloride of lime, rinsed in clean water, insufflated and submitted to pressure by rolling them under the arm, which stretches and enlarges them; then blown out quite tight, and fastened and dried.* See Cases **173** and **174**.

USES OF BLOOD.—Domestic economy and the industrial arts now use up a large quantity of the blood of animals which was formerly thrown away. The chief difficulty is that it is only possible to obtain it in quantity in extensive slaughter-houses, where many cattle are killed.

Its most important use is for making blood albumen, a manufacture which is carried on in this country and on the Continent, the product being cheaper than egg albumen. Blood is composed almost entirely of albuminous matter, of which 51·44 per cent. is contained in the clot which forms after the blood has stood, and 48·16 per cent. remains dissolved in the serum.

Two kinds of blood albumen are met with in commerce; ordinary, of which there are two or three qualities, and patent, which is transparent and soluble, and used for mordanting yarns and cloth for dyeing.

Ordinary albumen is prepared by adding a small quantity of spirits of turpentine to the serum and mixing it well, which bleaches it, and removes the grease. It is left to rest for 24 to 36 hours, and the serum which has become clear is separated from

* There is also a large and instructive display of all the economic uses of gut and bladder, in the Waste Products Collection, at the Bethnal Green Museum, formed by the author, of which a cheap special descriptive catalogue is published by the Science and Art Department.

the deposit. It is then dried on enamelled iron plates at from 50 to 57° Cent. for two hours, and the temperature lowered to 47½ or 49° Cent. for the space of 36 hours.

The patent or transparent albumen is prepared with diluted sulphuric and acetic acids and a further addition of turpentine, agitated for about an hour. It is then left for a day to rest, and the clear liquid drawn off, neutralised with ammonia, and dried. Ten pounds of serum will give about one pound of albumen.*

The colouring matter of the blood, hæmoglobin, may be obtained by evaporating its aqueous solution at a temperature below 100°; it then appears almost black, but resumes its red colour when dissolved in water. Coagulated blood is sold to calico printers for dyeing Turkey-red, and to chemical manufacturers for preparing red liquor for printers' use. About 6000 tons are estimated to be thus employed.

Dried blood serves to clarify wines, syrups, and other thick solutions. A very general use of blood is for manure, and it is one of the best fertilisers, equalling, in fact, powdered flesh. Blood for clarifying is of a good quality when it dissolves entirely in cold water, and when the solution of one pint of dry blood in ten of water, heated to boiling, produces an abundant scum and then leaves the liquor clear. An ounce of blood is usually sufficient to clarify a cask of wine. Dried blood used to be extensively employed on the sugar estates in the colonies to separate the scum and sediments in sugar boiling.

Attempts have frequently been made to utilise the blood of cattle as human food, but with little success, and yet it contains all the principles out of which tissues are formed, and hence must be eminently nutritious.

In Sweden they prepare a very good bread for the poor made with blood from the slaughter-houses and wheat flour. The blood

* The various blood albumens are shown in the Waste Products Collection at the Bethnal Green Museum.

of most of the domestic animals might thus be extensively utilised. In Denmark, Professor Panum, of the University of Copenhagen, has recently drawn attention to the amount of nutritious matter contained in blood and usually entirely lost, which can be preserved in forms suitable for food, as sausages, cakes, &c., mixed with fat, meal, sugar, salt, and a few spices.

We eat the blood of pigs and sheep in black puddings, but the blood of cattle and calves we throw aside as waste, an anomaly difficult to understand.

Recently a draught of blood has been recommended by physicians in cases of pulmonary phthisis, and in Paris and Chicago numbers of patients are said to resort to the slaughter-houses to drink the still fuming blood! This is somewhat like the practice of the African in Central Africa, who, Captain Burton tells us, severs one of the jugulars of a bullock and fastens upon it like a leech. This custom is common in Karagwah and the other northern kingdoms, and some tribes, like the Wanyeka near Mombasah, churn the blood with milk.

The Chinese seem to have great faith in blood. They use as medicine the dried blood of many birds and animals. The blood of the goat they consider a specific in pleurisy. They open the jugulars of the deer and by a long tube drink as much blood as the stomach will support. The coagulated blood of the rhinoceros is also used in Siam medicinally in case of inward hurts.

GLUE is an important commercial animal product which requires a passing notice. The quantity annually used in this country probably exceeds 10,000 tons, of the aggregate value of fully £500,000, as the prices range from £40 to £75 per ton. As a cementing substance glue is extensively used in every country.

Many refuse products are used in its manufacture. Animal skin in every form uncombined with tannin may be made into glue. The substances most largely and generally employed are the parings of hides, and skins from the tanneries and slaughter-houses known as "glue pieces," fleshings, calves' pates, &c., pelts from furriers, the hoofs and ears of horses, calves and sheep. The parings of ox and other thick hides make the strongest, and afford about 45 per cent. of glue.

Dried sinews, sloughs, (the core or bony support inside horns,) fresh bones, with other offal and garbage, are other raw materials used for making glue and size.

The commercial value of glue pieces ranges from £18 to £30, and when fleshed from £30 to £40 per ton. Sheep glue pieces and fleshings are of much less value, the price being from £10 to £15 per ton.

The process of manufacturing glue is as follows. The fresh fleshings, as well as the clippings of other skins which contain fresh lime, are steeped for several hours in water acidulated with sulphuric acid. Old fleshings, in which the lime is killed by becoming carbonate, are merely washed with water; and these with other glue-making materials are put into large open boilers, called "glue-pans," with water, and boiled for two or three hours by a naked fire, when glue is made, or by means of a steam-coil, when size is the product, until they are dissolved. The boiling gradually converts the "osseine" into gelatine, which is dissolved in the water. They are frequently stirred during this operation, in order that the fat may rise for collection. The liquid is then run off through a rough strainer into a tank, and allowed to settle for about half-an-hour, when it is either put into tubs and sent away as size, or it is allowed to set in wooden troughs, from which it is taken and cut up into blocks of glue about a foot square, which are subsequently further divided, by means of wire, into slabs and dried. The dry cakes are then dipped into hot water, and slightly rubbed with a brush to give them a gloss, and, lastly, stove-dried for sale. This furnishes the best and palest glue. The degree of concentration in making size is much less than that of glue, the point in the latter case being determined by the appearance of the cooked liquor upon a lump of alum.

The residue in the glue-pans is a mass of fibrous matter, called "skutch," which often contains enough fat to pay for another operation. The skutch is put into a boiler with enough sulphuric acid to dissolve the fibre (about 75 lbs. to a ton of skutch), and heated by high-pressure steam blown in it. Under this treatment the fibre dissolves, and so lets loose the fat, which rises to the surface.

Good glue should contain no specks, but be transparent and clear when held up to the light. The best glue swells without melting when immersed in cold water, and it resumes its former size on drying. The best method for use is to steep small pieces in cold water for twelve hours, then set it over a fire and gradually raise its temperature until it is all dissolved. Amber-coloured glue is that most esteemed by cabinet-makers.

Shreds or parings of vellum and parchment make an almost colourless glue; old gloves, rabbit skins, and such like, are also used for making size and gelatine.

Size is a weak solution of glue allowed to gelatinise. To transform glue into the gelatine of the shops it is simply necessary to dissolve it in water and allow it to settle. Clarifying agents are also used to destroy the last vestiges of colour.

By the use of gelatine, elastic moulds are made capable of reproducing with accuracy, and in a single piece, the most elaborately sculptured objects of exquisite finish and delicacy.

A gelatine is made from what is called "picker waste"—a picker is a band of buffalo hide used in driving the shuttles of power-looms. This gelatine or size is used for stiffening or dressing straw hats, silk, and other textile fabrics. The pieces cut off in making it are converted into gelatine for food purposes, but edible gelatine is also frequently made from sheep's trotters, old parchment, and waste pieces of glue.

There is a very large and fine collection of the glues and gelatines made in various countries shown in the Animal Products Collection, Bethnal Green, see also Cases 129 and 134.

HERD OF MIXED CATTLE IN AUSTRALIA.

Boiling down Cattle.—Sometimes, in large slaughtering establishments abroad, in Australia, or South America, the whole ox or sheep is sent to the melting-pot to be boiled down, for want of demand for its flesh or facilities for preserving it.

In such cases, as for instance in the vast establishment of Mr. J. H. Atkinson, Collingwood, near Liverpool, New South Wales, 70 to 100 men are employed boiling down sheep and cattle:—As soon as the ox is killed, he is lifted for skinning by machinery, and when the hide, head, hoofs, &c., are removed, the carcase is let down on a chopping-block running on a tramway; it is then cut into convenient-sized pieces, without the necessity of the men handling or lifting the meat, and the trolley chopping-block run on the rails to the other end of the building, where the boilers are. The meat is then lifted from the chopping-block into the boilers by means of endless chains with hooks attached, passing over sheaves, and driven by steam. The boilers are large steam-tight double cylinders, and capable of holding upwards of fifty bullocks at a time. When filled with meat, the orifice in the top of the boiler is closed, and the steam is let on at a pressure of 15 lbs. to the inch. In about seven hours the whole mass of meat and bone is reduced to a pulp. The steam is then condensed, and the tallow floats on the surface. On a tap being turned, it flows into the refining-pans; and when the refining is completed, by turning another tap, it runs into large shallow coolers. These are only about 3 inches deep, but very wide and long, in order that as great a surface as possible may be exposed to the air. When sufficiently cool, by turning other taps, the tallow is filled into casks alongside, and these are run by means of a tramway on to the weighing-machine, and thence to the rail for conveyance to Sydney. The mass of pulp to which both bone and flesh have by the steaming process been reduced, is then removed from the boilers by means of an opening near the bottom, fitted with a steam-tight door. It falls into a powerful press, also running on the tramways, and strong pressure being applied, a large quantity of highly concentrated soup is extracted; the flesh and bone having by the pressure been made into enormous solid cakes, the trolley-press is run into the piggery, and the greaves given to the pigs. The concentrated gravy or soup is then placed in a peculiarly constructed boiler, and reduced by evaporation to such a consistency that, when cold, it becomes solid, previously to which, however, it is run into bladders. It is, when cold, semi-transparent, of a rich reddish-brown colour, and sweet to the smell and taste, almost like confectionery. The first shipment from Sydney of this concentrated soup to England, was made in June 1862. An average bullock will yield about 20 lbs. weight of this portable soup.

The above account applies to cattle which are wholly boiled down. The prime portions of the best beasts, however, instead of being carried on the tramway to the boilers, are run off to the salting-house. The process there need not be described, further than that every particle of bone is extracted previous to the meat being salted. The leaner portions, not suitable for the casks, are cut into strips, and made into what is known as *charqui*, or *tasejo*, a South American name for dried or jerked beef. Each bullock will yield on an average about 100 lbs. of charqui, and the market for it is understood to be practically unlimited.

I need not go into the details of curing the hides, drying and smoking the tongues, extracting the oil from the hoofs, preparing the horns and leg-bones for the English market—or into the fellmongering, or sorting, washing, and scouring of the wool—for large numbers of sheep are slaughtered, as well as cattle. From the abundance of water, however, all these processes are carried on with a degree of cleanliness and an absence of offensive smells most surprising.

COW-HAIR.—Besides the cow-hair collected at home, about 3000 tons are imported annually, chiefly from Germany and France. Russia ships about 64 tons.

Cow-hair used to be extensively employed for mixing with

mortar for building purposes. It is now made into felt for roofing, and for clothing boilers and pipes of steam-engines. It is used for twisting into rope, and stuffing sofas and chair cushions. In some parts of Germany carpets are made of cow-hair. The demand for cheapness has stimulated the makers of inferior textiles and blanketing to mix cow-hair with wool. It is also used in the fabrication of horse-cloths and railway rugs, and ladies may be interested in knowing that the so-called cheap sealskins are made in the north of England from hair that used to go to the plasterers to bind their mortar. Wet cow-hair is sold at the tanneries for about 2*s.* 6*d.* a bushel, and is washed and dried and the lime beaten out. White hair is worth nearly double the price of coloured hair. Plasterer's hair ranges from £5 10*s.* to £8 10*s.* the ton, washed cow-hair £10 to £11 the ton.*

* The uses of cow-hair for felt and other purposes are shown both in the Animal Products Collection (Case 95) and in the Waste Products Collection at Bethnal Green Museum.

CHAPTER V.

THE DEER AND ANTELOPE TRIBES.

Having considered the Bovines and their economic uses, we are next led to investigate another class of Ruminants—the Deer tribe, and to consider what products they furnish for the use of man. Although not very common in civilised Europe, in many regions they are of essential importance, both for food and other purposes. The horns of these animals are first alluded to, their specialities and trade uses; then the commerce and applications of deer skins receive notice; after that the various species of deer of any importance in an economic point of view are described, including the reindeer, musk-deer and its product; we then pass on to the Antelope tribe, those of any special economic interest, as the eland, the spring-bok, and the gnu being noticed.

THE DEER TRIBE.—Besides their flesh, which is eaten as food when obtainable, the products which the deer tribe supply to commerce are chiefly antlers or horns, and skins. We shall speak first of the horns.

There are several species of deer. For our purpose these may be divided into two groups, of which one includes those with antlers more or less flattened, the other those with rounded antlers.

The elk (*Alces*) is the most characteristic species of the first group. The reindeer differs from the rest of the genus in the presence of antlers in both sexes, and in the great development of the brow antlers. The English park or fallow deer (*Cervus dama*, *Dama platyceros*) is referable to the flat horned group.

The number of stags and hinds in Scotland is rather more than 10,000.*

* In the Bethnal Green Collection there are mounted heads of the fallow deer, male and female, of the roe, male and female, the red deer, male and emale, a fine head and antlers of the wapiti (*Cervus canadensis*), of the roebuck,

Deer-horn produces a great quantity of gelatine by decoction, and the raspings of deer's horn are occasionally employed in domestic economy to furnish what is supposed to be a nourishing jelly. The waste pieces are sometimes boiled down for size in the cloth-making districts. Submitted to the action of heat the product is the same as that of most animal substances. It used to be largely employed for the produce of ammonia.

HEAD AND ANTLERS OF THE ARCTIC REINDEER.

The first year the stag has properly no horns, but only a kind of corneous excrescence, short, rough, and covered, with a thin, hairy

an unusually fine head and antlers of the *Cervus axis*, two heads with antlers of the reindeer, and varieties of deer horns from India, Siam, &c.; also feet of the elk or moose formed into paper racks or pockets in Case **96**; a stuffed head of the eland, and horns of the waterbok and various other antelopes.

skin; the second year the horns are single and straight; the third year they have two antlers; the fourth, three; the fifth, four; and the sixth, five. When arrived at the sixth year the antlers do not always increase, and though the number may amount to six or seven on each side, the stag's age is then estimated rather by the size and thickness of the branch that sustains them than from their number. The proportional length, direction, and curvature of the antlers vary, and it oftens happens that there is one more or less on the one side, than on the other; the horns also become larger, the superficial furrows more marked, and the burr is more projecting. Nothwithstanding their magnitude, those horns are annually shed in the spring of the year, and succeeded

STAGES OF GROWTH.

by new ones. A full grown stag's horn frequently weighs 24 lbs., and the whole of this immense mass of true bone is produced in about ten weeks.

The form of the horns differs at different ages; but it is not so easy to tell the age of a stag by its horns.

A correspondent in *Land and Water* thus describes the stages of growth. Eight or nine months after birth the horns appear as nearly straight branches, growing to the length of from six to twelve inches. This single horn being cast, the two-year-old stag sets up his second pair, on which one side prong appears a few inches above the point of junction with the head. In the third year each horn acquires two ends. The lower prong increases in length, and is developed somewhat lower down on the horn. The upper prong is somewhat smaller than this. The third end preserves the character of the original single horn. In the next year the stag sets up four points on each horn, the lower prong drawing still nearer to the base, and the original horn breaking out into two points. The next stage is the development of points between the eye-end and the middle-end, which are materially smaller. In the next stage the two upper points form a crown.

The stag is then called a stag of twelve points, or crown stag, or, as we say in Scotland, a Royal Stag. The cut on p. 179 shows the progressive development up to twelve points, as given in Winckell's "Manual for Sportsmen," and Altum's "Forest Zoology."

Up to the eighth or ninth year the density of the horns increases, and from that to the twelfth year the horns are in greatest perfection. The number of tiers or branches of the horn varies according to the age of the animal. At the present day the oldest stags in Scotland seldom present more than ten or twelve "points." There is a head still preserved at Mauritzberg which presents the enormous number of sixty-six points. It was killed by the first king of Prussia, and presented by that monarch to Augustus, Elector of Saxony and King of Poland. In the Collection at the Chateau of Wohrad, the hunting residence of the Lordship of Fauenberg, there are one hundred and nine stags' heads, of which only seventeen are under fourteen points.

There are occasionally curious contortions in the horns. Mr. A. Murray, in the "Edinburgh New Philosophical Journal," describes those of an old reindeer, but small in size and with small horns. The horns have met with a distortion, by which they have a curious bend in the middle, as shown in this figure. The cause, whatever it may have been, has affected them both equally, which is not usually the case where horns are distorted. It may be that the poor animal, when its horns were still soft and young,

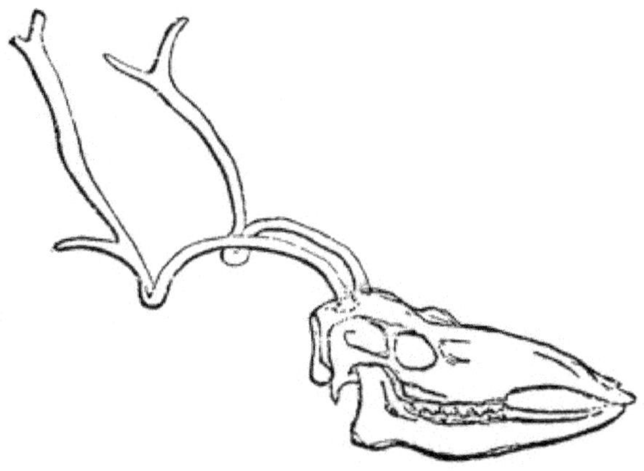

got entangled among brushwood; and that here is the silent evidence of long struggles on the part of the animal, and of perhaps days of famine before it succeeded in freeing itself from the bonds which held it. Or it may be merely a distortion consequent upon the old age of the animal, for we often find the horns in old deer stunted and distorted, although it is not usual to find them so symmetrically disfigured.

The horns of the deer, more properly called "antlers," are solid processes from the frontal bone, and possess the chemical and physical properties of true bone. After being sawn and filed to the required shape, the exterior is left in its rough and natural state, which, besides being ornamental, is well adapted for the handles of knives and instruments requiring a firm grasp. In the German States, very pretty and delicate objects are carved from this material.

Buck-horn is principally used now for handles of carving-forks and knives, pocket knives, solid or in scales, and fancy articles. It should be chosen with hard stems, like ivory, yellow and furrowed on the exterior, but of a heavy white in the interior.

In Holland, Hamburg, and other parts of the Continent, buck-horn is more worked up into useful articles than in this country. At the various International Exhibitions many large fancy articles have been shown made of stag-horn, such as umbrella-stands, chandeliers with many lights, &c. In Switzerland the stag's-horn is manufactured into a variety of ornamental articles for personal wear, such as brooches, pins, bracelets, and many other things.

The stag-horn used in Sheffield for cutlery purposes is chiefly supplied from India and Ceylon, the shipments reaching about 400 tons annually; and another 100 tons is obtained from Europe, from our own deer parks, and other sources.

Of all the deer family the elk (not yet extinct in Germany and Scandinavia) has the largest horns. Those of the species inhabiting the hills and plains of India and Ceylon are also very heavy. From Southern Russia we have deer horns lighter than the Indian, but heavier than the German, and the antlers of the reindeer are lighter and more porous than any others. The surface is nearly white and smooth, and consequently they are much less valuable.

The number of horns to the ton necessarily varies according to their size, ranging from 700 to 2,000. Taking the imports from the East at 350 tons, and assuming an average number of 1,400 horns to the ton, it would appear that the "fall" as it is termed of 245,000 head of deer is annually collected there for the use of our manufacturers. The Greenland Company of Denmark receives about 30,000 reindeer's horns or antlers yearly.

The fine elk horns exported from Germany are getting scarcer every year, for the forests decrease, and the native cutlery manufacture extends.

The cut on the next page shows a pair of antlers in the Museum of the College of Surgeons, which have a span of 35 inches, and are 7 inches apart at the base.

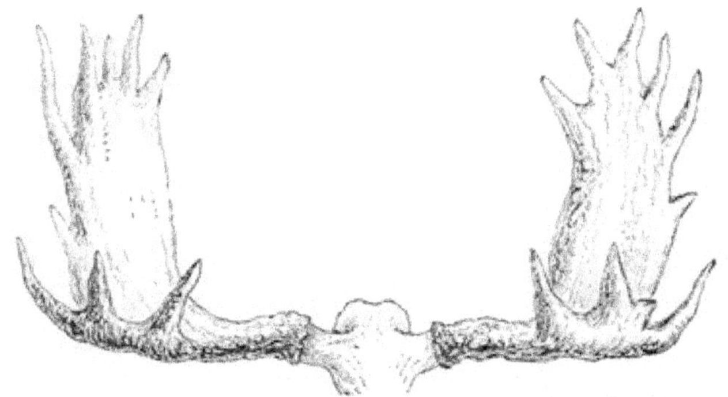

ANTLERS OF ELK OR MOOSE (*Alces machlis*, Ogilby).

The value of the different kinds of stag-horn is in the following order: German, Italian, Russian, and English. The total imports of stag-horn into Britain are estimated at about 700 tons yearly.

Deer-horns form a large item of commerce in many countries. About 26,000 pairs are annually shipped from Siam. In 1869, 1,433 cwt. of deer-horns were shipped from Ceylon, valued at £2 per cwt. They consist of the horns of the Rusa deer (*Cervus hippelaphus*, Ogilby) and of the spotted deer (*Axis maculata*). In Cashmere the antlers of the Hungul or Persian deer (*Cervus Wallichii*) shed early in spring, are picked up in the forests by the villagers, and form an article of export traffic with Ladak.

DEER SKINS.—There is a large commerce in deer skins in various countries, although we do not receive very many here, probably about 80,000 are imported annually, for the most part undressed, which are tanned into leather for various purposes. About 1,000 cwt. are annually collected in Slavonia, about one third of which are exported. From Guatemala 35,171 skins were exported in 1874, valued as low as 4*d.* per pound. From Venezuela about 300,000 skins of deer and other animals are annually exported. In the Museum Collection are skins of the reindeer and other deer. Case **97**.

Deer are found in great abundance in every part of the Minnesota Territory and other portions of the Western States of America and the settlers during the winter season are usually well supplied with venison. The best haunches are worth from 4*d.* to 5*d.* per pound, and when properly cooked are savoury.

The skins are valuable for many purposes, and are neatly dressed by the Indians, and manufactured into a great variety of useful and ornamental articles. Deer skins vary in price per pound, according to weight. An ordinary skin, weighing from eight to ten pounds, is worth from 5*d.* to 7*d.* per pound, while one weighing twelve or fifteen pounds is worth 1*s.* per pound. They average about 7½*d.*

In 1863 the Hudson's Bay Company imported 10,751 deerskins; but the average is seldom more than 3,000 a year.

From 2,000 to 4,000 packs or bundles of deer skins reach New Orleans yearly by the Mississippi, which are valued at about £4 the pack. Gloves and gauntlets of buck-skin are made in the United States, of the white tanned skins of the common deer. Deer skin leggings or breeches made of deer skin are much liked and used by the hunters in North West America.

In the small Indian village of Lorette in Canada no less than 2,500 deer skins, besides a large number of those of other wild animals, are worked up annually into winter shoes and mocassins, and as many as 20,000 pairs are made from that number, selling wholesale at an average of from one to two dollars per pair; 1,000 pair of raquettes or snow shoes, selling at from three to six dollars a pair; 300 Indian sleds or toboganes selling at from one to two dollars a pair; besides a quantity of ornamental articles of considerable value—thus yielding a return of upwards of £7,000 to the hunters and natives of this one village for their year's industry.

Deer hair when it can be obtained is used for stuffing saddles, for which it is a good material. Case **96** contains samples of deer hair and fancy Indian embroidery work made with it in Canada, slippers, baskets, &c.

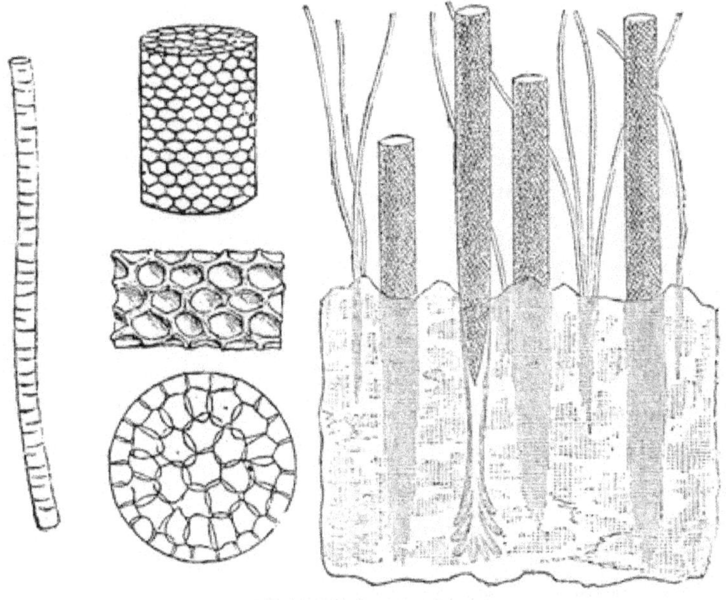

DEER HAIR MAGNIFIED.

It has been made known by Professor Busk, that the hair of the deer tribe is peculiar, being almost entirely cellular; and the hair has been described and figured by Dr. Inman in an able paper, "On the Natural History and Microscopic Character of Hair," published in the Proceedings of the Literary and Philosophical Society of Liverpool. The subject here figured is the skin and hair of one of the North American reindeer; but the structure seems to be the same in all deer—in the moose, the red-deer, roe-deer, musk-deer, &c., but not in the antelopes.

The figure on the right hand represents a somewhat magnified portion of the skin, with both kinds of hair issuing from it; the left-hand figure represents a more highly magnified small hair; the upper centre figure shows a highly magnified portion of the large hair; the lower centre figure, a transverse section of this; and the middle centre, a longitudinal section.

Dr. Inman says: "In the deer the cells are so numerous as to occupy the whole of the body of the hair, and so irregular that no particular place of subdivision can be traced." And his figure quite corresponds with this, the cells being there shown as amorphous; but it will be seen from the above figure that they are truly polygonal—for the most part hexagonal. There are very distinct septa and lines of separation.

"It is held by physiologists that both these kinds of hair are modelled on the same plan, viz., that of a cellular interior surrounded by a horny cortical exterior, and that the difference in texture arises from the difference in the extent of development of the internal cellular pith, or of the external cortical covering: in the one extreme forming the soft hair of the deer; in the other, the hard bristle of the sow. This view recommends itself by its simplicity, and the unity of the *modus operandi;* but although it may be correct, so far as it goes, it does not explain the whole of the phenomena. For example, it does not explain why the hairs, where the horny covering predominates, are imbricated, while those which are cellular are not; and it is to be observed that there is a want of transition between the two characters of hair, which certainly is opposed to a common mode of development. If it were the same, we ought to find hairs exhibiting all the gradations of passage between the two extremes, which we do not. Furthermore, they appear to be designed for different purposes. Speaking in a general way, the horny or bristly hair is characteristic either of carnivorous animals, who have a greater supply of caloric than vegetable feeders, or of graminivorous animals inhabiting warm climates; while the cellular hairs in question are confined to the deer tribe, most of whom inhabit cold climates. It has usually been said, that the fine hair found at the roots of the coarser hair in these animals is an additional provision of nature for the warmth of the animal. It rather appears to me,

DEER IN THE JUNGLE,

that, in the deer at least, it is the larger cellular hairs which have been added for this purpose (no one can look at them, I think, without seeing how admirably they are adapted for this), and that the horny hairs, whose office may possibly be as much that of a regulator of temperature as of a heating apparatus, are the normal hairs of the animal reduced to the smallest dimensions. If these two kinds of hair have distinct functions, their mode of development may also possess distinctive characters. We see that their roots extend to very different depths in the skin, and although we know that the hair is a mere appendage of the skin, produced by its involution or evolution, it may be that by drawing more of its substance from one layer than from another, the differences in its appearance which we have been considering are produced." *

Deers' sinews, dried, are esteemed a great food dainty in China, and some other eastern countries. The hoofs grated are also used medicinally to cure wounds.

Some half dozen species of Asiatic deer, according to Dr. Sclater, constitute the Rusine group, of which the Sambur (*Cervus Aristotelis*) may be taken as the type. Their antlers have but three points, are comparatively short in the beam; but especially in Ceylon these attain an immense thickness. Mr. Bates states that he has killed bucks carrying heads which measured eight inches in circumference at the burrs.

The flesh of the Rusa deer is considered a delicate morsel in Java and Borneo.

The Virginian deer (*Cariacus Virginianus*) is still found in large numbers in the unsettled parts of North America, and is an animal of great importance to the Indians, who appreciate and cure its flesh for winter provision.

THE MOOSE OR ELK (*Cervus Alces*, Richardson; *Alces Americanus*, Baird; *Alces machlis*, Ogilby), inhabits the northern part of both continents. This deer is in size rather larger than a horse.

* Andrew Murray, Esq., in the "Edinburgh New Philosophical Journal," New Series, April, 1856.

Its flesh is more relished by the Indians and persons resident in the fur countries, than that of any other animal. It bears a greater resemblance in its flavour to beef than to venison. It is said that the external fat is soft, like that of a breast of mutton, and when put into a bladder is as fine as marrow. In this it differs from all other species of deer, of which the external fat is hard.

The tough skin of the elk has been put to various uses. In Sweden a regiment was clothed with waistcoats made of this material, which was so thick as to resist a musket ball. When made into breeches, a pair of them among the peasantry of former days, went as a legacy through several generations.

A buck in its grease will weigh as much as 800 lbs., without the offal. When in good condition, the flesh is sweet and tender, and is highly esteemed as an article of food; but should the animal be poor, or have been subject to violent exertion previously to death, the meat is scarcely eatable. The *moufle*, or loose covering of the nose, is considered by epicures the greatest delicacy of the north-west, contesting the palm with bear's paw, beaver tail, reindeer tongue, buffalo boss, and sheep ribs. The Indians sometimes snare the moose; and in the spring, when the action of the sun has formed a thick crust upon the snow, they drive them into drifts, and spear them in numbers. It is not a gregarious animal, and to hunt it requires more skill than is necessary in the pursuit of either reindeer or buffalo. The moose furnishes an excellent hide for moccassins and snow shoes in America, the best skin is from the bull moose in October, and usually sells for about sixteen shillings.

The uses to which the various parts of the moose are put are many. The hide supplies parchment, leather, lines, and cords; the sinews yield thread and glue; the horns serve for handles to knives and awls, as well as to make spoons of; the shank bones are employed as tools to dress leather with; and with a particular portion of the hair, when dyed, the Indian women embroider garments.

To make leather and parchment, the hide is first divested of the hair by scraping, and all pieces of raw flesh being cut away, if then washed, stretched, and dried, it become parchment. In converting this into leather a further process of steeping, scraping, rubbing, and smearing with brains has to be gone through, after which it is stretched and dried, and then smoked over a fire of rotten wood, which imparts a lively yellow colour to it. The article is then ready for service. Of parchment, as such, the Chipewyans make little use, but the residents avail themselves of it, in place of glass for windows, for constructing the sides of dog-carioles, and for making glue. The leather is serviceable in a variety of ways, but is principally made up into tents and articles of clothing, and in the fabrication of dog-harnesses and fine cords, wallets, &c. The capotes, gowns, firebags, mittens, moccasins, and trowsers made of it are often richly ornamented with quills and beads, and when new, look very neat and becoming. The best Indian dressers of leather in the Canadian dominion are the Slave Lake Chipewyans and those of the Liard's River.

The lines and cords are of various sizes, the largest being used for sled lines and pack-cords; the smaller answer for lacing snow-shoes and other purposes. In order to make sled lines pliant—a very necessary quality when the temperature is 40 deg. or 50 deg. below zero Fahrt.—the cord is first soaked in fat fish liquor, it is then dried in the frost, and afterwards rubbed by hauling it through the eye of an axe; to complete the operation it is well greased, and any hard lumps masticated until they become soft, by which process a line is produced of great strength and pliancy, and which is not liable to crack in the most severe cold.

To obtain thread, the fibres of the sinews are separated and twisted into the required sizes. The moose furnishes the best quality of this article, which is used by the natives to sew both leather and cloth, to make rabbit snares, and to weave into fishing nets. Sinews can be boiled down into an excellent glue or size.

In mounting knives and awls with horn, lead, copper, and iron are used for inlaying, and rather handsome articles are sometimes produced. The making of spoons, tipping of arrows, carving of fish-hooks, stuffing of dog-collars, and embroidering with hair need no particular comment. [Mr. B. R. Ross, of the Hudson's Bay Company, in the "Technologist," vol. ii., p. 260].

THE REINDEER (*Tarandus arcticus*, Rich.), occupies a band of country fringing the polar zone in Northern Europe and Asia.

THE REINDEER.

It replaces the horse and the ox in a climate where these animals could not be utilised. In the Russian empire there are computed to be about one million, of which 680,000 are in Europe, chiefly in the governments of Archangel, Vologda, and Perm. In Siberia there are more than 300,000. The Koraks of Eastern Siberia who are nomads, and live in skin tents like the Tongouses, have immense herds of reindeer, some tribes own not less than 15,000. These supply their food, clothing, and means of transport. Their gut forms an excellent twine, and their bones serve to make various tools and arms, and enter into the formation of their sledges and vehicles which are often of elegant construction. The pride of the Laplanders is also to have large herds of reindeer for their sledges. They drink their milk and make cheese of it; they clothe themselves with the skins, and eat their flesh, which is good.

In Norway, in 1865, there were stated to be 102,000 reindeer,—of which about 60,000 were in Finmark. The possession of 300 reindeer constitutes the independence of a family; 500 riches. Some possess as many as 1,000. In October a fifth or a sixth are killed for food. A fine reindeer will sometimes yield 120 lbs. of meat and 40 of tallow.

ESQUIMAUX DOGS KILLING A REINDEER.

In America the reindeer is confined almost entirely to the "Barren-grounds" (whence it takes its common name), the north-eastern corner of North America, along the Polar Sea, bounded to the west by Great Slave, Athabasca, Wollaston, and Deer lakes, and the Copper Mine River, and to the south by Churchill river. Here the Barren-ground reindeer graze by thousands, accompanied by the musk ox, another characteristic inhabitant. Being so plentiful it is termed the common deer by the hunters, just as the *Cervus Virginianus* bears this name in the United States. There would seem to be two varieties, if not distinct species, of this animal met with; one termed the Strong-wood reindeer, which inhabits the thickly-wooded parts of the district, particularly among and in the vicinity of the mountain-ranges, where they are of very large size. Though smaller than the moose, these deer are of considerable bulk, and weigh up to 300 lbs. In most particulars they resemble the Barren-ground species, differing from it in the following points:—smaller horns, darker colour, larger size, not being so gregarious, and not migrating. The only hides serviceable for converting into leather are those of animals killed early in the winter, which, when subjected to the process already described, and bleached in the frost, instead of being smoked, furnish a most beautiful, even, and white leather, which is used for shoe-tops, and is embroidered with quills and silk.

The Barren-ground reindeer, during the summer and spring months, frequent the barren plains lying between the wooded country and the shores of Hudson's Bay and the Arctic Sea. Their migrations, which are performed with wonderful regularity, are as follows: They leave the shelter of the woods in the end of March and beginning of April, and resort to the plains where they feed on various kinds of lichens and mosses, gradually moving northward until they reach the coast, where they bring forth their young in the beginning of June; in July they begin to retire from the sea-board, and in October rest on the edge of the wood, where they remain during the cold of winter. The horns of these deer are much varied in shape, scarcely any two animals having them precisely alike. The old males shed theirs in April, and the gravid females in May. Their hair falls in July, but begins to loosen in May. Their new coat is darkish brown and short; but it gradually lengthens, and becomes lighter in colour, until it assumes the slate-grey tint of winter. The flesh, when in prime condition, is very sweet, but bucks, when in season, have their fat strongly impregnated with the flavour of garlic, which, indeed, is always present more or less.

The shortness of the hair of the Caribou or Barren-ground reindeer, and the lightness of the skin when properly dressed, render it the most appropriate article for winter clothing in high latitudes. The skins of the young deer make the best dresses; and the animals should be killed for that purpose in August, as after that month the hair becomes long and brittle. They are so drilled into holes by the larvæ of the gad-fly that eight or ten skins are required to make a suit of clothes for a grown person. But the skins are so impervious to cold that, with the addition of a blanket of the same material, any person may bivouac in the snow with safety, and even with comfort in the most intense cold of an arctic winter's night.

Every part of the carcase serves the natives for food. The hunter breaks the leg of a recently slaughtered deer and swallows the marrow, still warm, with avidity; the kidneys, and other parts of the intestines are also eaten raw; the colon, or large gut, when roasted or boiled, with all its fatty appendages, is one of the most savoury dishes that can be offered, either to Indian or white settler in North America. The stomach, with its contents of lichens and other vegetables, is also eaten—the latter substances being much more easily digested after they have partially undergone that process by the gastric juice of a ruminating animal. Some Indians and Canadians leave this savoury mixture to ferment or season for a few days before they eat it. The blood, if mixed in proper proportion with fat meat, and cooked with some nicety, forms a rich and highly nutritious soup. After all the flesh is consumed, the bones are pounded, and a large quantity of marrow extracted by boiling; this is employed in preparing pemmican.

Reindeer's tongues are much liked by many in this country, large quantities being imported annually from Russia. They are snow cured, no salt whatever being used, the mildness and richness of flavour in the meat is preserved, and they are rendered extremely acceptable to refined palates.

In North America the reindeer supplies the Chipewyans, Copper Indians, Yellow Knife, Dog Rib, and Slave Indians, with food, who would be totally unable to inhabit their barren lands were it not for the immense herds of this deer that exist there. Of the horns they form their fish-spears and hooks; and previously to the introduction of iron by the traders, ice chisels and various other utensils were made of them. In dressing the skins, the shinbone split longitudinally, is used for the purpose of scraping off the hair, after it has been repeatedly moistened and rubbed. The skins are then smeared with the brains of the animal until these acquire a soft, spongy character, and lastly are suspended over a fire made of rotten wood, until thoroughly impregnated with the smoke. This last mentioned process imparts a peculiar odour to the leather, and has the effect of preventing its becoming so hard, after being wet, as it would otherwise be. The skins thus dressed are used as winter clothing; and 60 or 70 sewn together will make a covering for a tent sufficient for the residence of a large family. Their clothing for winter is made out of fawn skins, dressed with the hair on, and consists of capotes, gowns, shirts, leggings, mittens, socks, and robes, which are warm, and when new, nice looking. Hides which are so much perforated by the larvæ of the *œstrus* as to be unfit for any other purpose, are converted into *babiche*, to make which the skin is first divested of hair and all fleshy matter; it is then with a knife cut into the desired thickness, the operation beginning in the centre of the skin. There are two sizes of this article: the larger being used for barring sleds and for the foot-lacing of snow-shoes; the smaller as a species of thread for sewing leather, for the fine netting of snow-shoes, and for lacing, fishing, and beaver nets. [B. R. Ross.]

The undressed hide, after the hair is taken off, is cut into thongs of various thickness, which are twisted into deer snares, bow-strings, net-lines, and in fact, supply all the purposes of rope. The finer thongs are used in the manufacture of fishing nets, or in making snow shoes, while the tendons of the dorsal muscles are split into fine and excellent sewing thread. In some instances the skin is so finely dressed that it equals chamois leather.

The caribou travel in herds, varying in number from eight or ten to one hundred thousand.

The number of reindeer killed annually in South Greenland has much decreased of late years. It used to average over 18,000 head. In the five years from 1840 to 1845, it was calculated that on an average, 16,000 were killed yearly, from 1851 to 1855, only about half this number were killed, and it is probably now less. The Greenland Company only sold 637 skins in 1861. When hunting, the Greenlanders merely use the flesh as they want it for their daily meals, and let the rest lie without putting it to any use. The animals are generally shot a long way up in the country, and consequently from the difficulty of conveying them to their huts, a great part is often left to waste on the ground. The flesh is often exchanged by the Laplanders for flour. The gloves which are sold as real Swedish gloves are obtained from the young whose mother has died before giving birth, which is rare. It takes three of these skins to make two pair of gloves. Each skin costs in Sweden 3*s.* or 4*s.*

Dall, in his work on Alaska, states that at one village he counted 1072 bunches of fresh skins of the reindeer fawn hanging up to dry. As each bunch contained four skins, or enough to make a "parka," this would (he adds) give a total of nearly 4300 of these little creatures which had been killed in two months.

The Laplanders also use part of the entrails for making a kind of thread or cord which is difficult to break, and which is much sought for in England. Under the name of wire of Lapland, a fine slender substance was formerly prepared by the Laplanders, in thickness and appearance resembling our silver wire. It was made of the sinews of the reindeer, reduced to the finest filaments and then drawn through melted tin for the purpose of coating it with that metal.

The milk of the reindeer constitutes the food of the wandering Laplanders during the great part of the year.

The stag, or red deer (*Cervus Elaphus*), the fallow deer (*Dama vulgaris*), and the roebuck (*Capreolus capræa*), are hunted for their flesh, and their skin and horns are applied to economic uses, similar to those already described.

MUSK DEER (*Moschus moschiferus*, Lin.).—This animal, which inhabits the mountains of China, Thibet, and Siberia, is the source of the musk of commerce, a perfume which is in demand in nearly

MUSK DEER.

every part of the civilized world. There is a good stuffed specimen of the animal in the Bethnal Green Museum. Musk is one of the most permanent of all odours, and heightens the aroma of many other perfumes. In the animal kingdom there are several other animals pervaded with the musky odour, such as the musk rats of India, Europe, and America, the musk ox and the buffalo, the alligator and crocodile, some birds and insects.

In most of the Hill States the musk deer is considered as royal property. In some the Rajahs keep men purposely to hunt it, and in Gurhwal a fine is imposed on any who are known to have sold a musk pod to a stranger, the Rajah receiving them in lieu of rent.

It is only from the follicles or pouches of the male deer that the musk is obtained, and in some adult males the pod may contain as much as two ounces. An ounce is about an average; and as many of the deer are killed when young, the pods brought to market may be taken to average all round only about half an ounce. A single grain of musk will fill the air of a large apart-

HUNTING-PIECE, FROM A CHINESE DRAWING, IN WHICH DEER, LEOPARD, AND MUSK-DEER CAN BE SEEN.

ment with a sensible impregnation for many years without its weight being perceptibly diminished, and one part can communicate its odour to 3,000 parts of an inodorous powder. Our imports of musk range from 10,000 to 17,000 ounces yearly. The exports from British India in 1875 were 7,403 ounces.

The musk bags or sacs, after the grain musk has been extracted, are used by perfumers to prepare essence of musk.*

There are two commercial kinds of this musk, named after the countries where they are obtained, the Tonquin or Thibet musk,

* For these illustrations we are indebted to Rimmel's "History of Perfumes."

imported from India, and the Kabardin or Siberian musk which comes through St. Petersburgh or by the way of China. Although the musky odour penetrates the whole animal, its flesh is said to be eaten by the natives. The skin is manufactured into furs and leather.

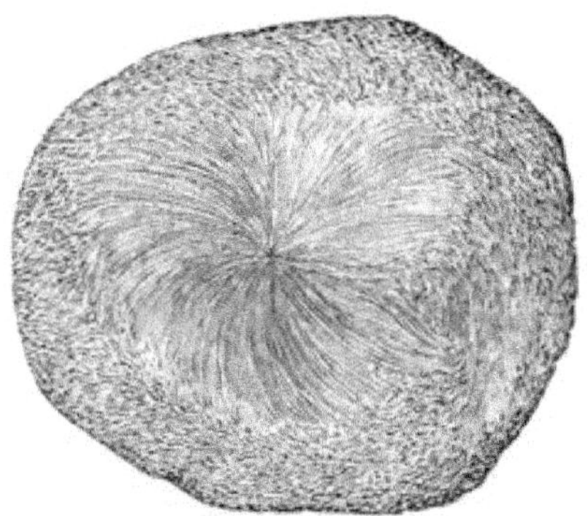

MUSK POD, NATURAL SIZE.

THE ANTELOPE TRIBE.—These animals are confined to the old world, but in no other part are so many varieties of the family seen as in South Africa, over large districts of which they wander. Not less than twenty-seven species, from the stately blackbok down to the diminutive blaabok or pigmy antelope, many of which, however, are found in other parts of the continent, are reckoned south of 20° of S. latitude; of these the largest, the eland, is still met with in the western parts of Natal, and more plentifully in the Zulu country, the Transvaal territory, the Kalihari, the Betchouanaland and the Ngami regions. It was once very common in every part of the Cape colony, as the numerous localities called by its name testify.

THE ELAND (*Oreas Canna, Boselaphus oreas*, Gray).—Among the known species of antelopes, which are not less than 80 in number, there are none more imposing from its size or more interesting in an economic point of view, than the Eland. Sir Cornwallis Harris speaking of it says :—" In shape and general aspect, he resembles a Guzerat ox, not unfrequently attaining the height of nineteen hands at the withers, and absolutely weighing from fifteen hundred to two thousand pounds! By all classes in Africa the flesh of the eland is deservedly esteemed over that of any other animal. Both in grain and colour it resembles beef, but is far better tasted and more delicate, possessing a fine game flavour, and exhibiting the most tempting looking layers of fat and lean, the surprising quantity of the former ingredient with

WHITE-EARED ANTELOPE (*A. leucotes*), MALE, CENTRAL AFRICA.

which it is interlarded exceeding that of any other game quadruped with which I am acquainted. The venison fairly melts in the mouth, and as for the brisket, that is absolutely a cut for a monarch. During the greater part of our journey it was to the flesh of this goodly beast that we principally looked for our daily rations, both on account of its vast superiority over all other wild flesh, and from the circumstance of its being obtainable in larger quantities with comparatively less labour."

Lichtenstein, in his "African Travels," states that the meat is cut in pieces on the spot, salted and packed in the skins, and some of it is smoked. The great muscles of the thigh, smoked, is more particularly esteemed; these are cut out at their whole length, and from the resemblance they bear to bullocks' tongues are called by the Dutch thigh-tongues. They are eaten raw with bread and butter, cut into very thin slices. The taste of eland's flesh when eaten fresh resembles beef. The skins are much esteemed for making leather, and the horns are formed into tobacco pipes. The eland has been introduced and acclimatised in Europe.

THE SPRINGBOK (*Gazella euchore*, *Antidorcas euchore*, Gray).—This antelope is numerous between lat. 27° and 32° S. It is gregarious and found in flocks of many thousands, and is common throughout Syria, Egypt, Morocco, Barbary, Greece, and the Holy Land. The flesh is in universal esteem among all the people. It is fine flavoured when fat, and of a delicate taste. The Moors keep it in confinement as a pet. Jules Gerard tells us that its dung dried in the sun and reduced to powder, gives a very agreeable flavour and odour to the tobacco smoked in Algiers. The horns of the springbok are shaped like a jews-harp, are twelve to fifteen inches long, and annulated to within three inches of the tips.

The bok much resembles the English deer, with short, straight, or curved horns, instead of antlers. Of these there are a considerable variety, and a very great number in every district of South Africa. Some, as the bushbok (*Tragelaphus sylvaticus*) have comparatively short necks and legs, with a stout, closely-knit frame, standing scarcely higher than sheep; others, as the rheebok (*Eleotragus capreolus*, Gray), are tall and stately, with long legs, long straight necks, and straight horns, with an exterior twist at the roots. The flesh of the bok (pronounced like our English word buck) makes very fine venison when properly dressed, and the legs and shoulders of the animal are much esteemed as a relish when dried down into *biltong*, a most convenient and palatable article of diet, perfectly familiar to the colonists. In this form it can be kept almost any length of time, and has frequently been brought to England. It is extremely nourishing and digestible, and can often be taken by individuals when other food is rejected by the stomach.*

Superb karosses or cloaks are made and worn by the Kafir chiefs, composed of the skins of antelopes and other wild animals neatly sewn together, among which the leopard skin, jackal, and fox skin, seem to predominate. They are much valued for the variety and beauty of the skins of which they are composed, as well as for the neatness of the workmanship, all the pieces being sewed together with sinew. Hundreds, if not thousands, of these are purchased annually by traders and sent into the markets of the Cape Colony for sale. A good one will fetch £8 or £9.

In former times the occasional migrations of the springboks into the settled districts of the Cape Colony used to be looked upon with dread. The number of these animals sometimes seen in the Karroo Plains, within a compass of fifty miles, has been computed to be at least 100,000. It is scarcely possible for a person passing over the extensive tracts of the interior and admiring this elegant antelope, thinly scattered over the plains and bounding in playful innocence, to figure to himself that these ornaments of the desert can often become as destructive as the locusts themselves. The incredible number which sometimes pour in from the north, during protracted droughts, distress the farmer inconceivably. Dr. Livingstone states that in their migrations, when first they cross the colonial boundary, they are said to exceed forty thousand in number.

He also mentions that great numbers of wild animals, gnus, koodoos, zebras, &c., die from pleuro-pneumonia, although the mortality produces no sensible diminution in the quantity of game. He inculcates, therefore, caution, for "when the flesh of animals that have died from the disease is eaten it causes a malignant carbuncle; and when this appears over any important organ it proves rapidly fatal. It is more especially dangerous

* Silver's "Handbook for South Africa."

over the pit of the stomach. The effects of the poison have been experienced by missionaries who have partaken of food not visibly affected by the disease. Many of the Bakwains who persisted in devouring the flesh of animals which had perished from the distemper died in consequence. The virus is destroyed neither by boiling nor roasting. This fact, of which we have had innumerable examples, shows the superiority of experiments on a large scale to those of physiologists in the laboratory, for a well known physician of Paris, after careful investigation, considered that the virus was completely neutralized by boiling."

BASTARD GEMSBOK (*Antilope leucophæa*, Pallas).

THE GNU.—There are two well defined species of this antelope. 1. The common black gnu or kokoon of the Bechuanas (*Catoblepas* [*Connochetes*] *Gnu*), which is found in the plains of the Free State and the Transvaal, with horns bent forward and downwards, points bent at an acute angle upward. Their length being 26 inches.

2. The brindled gnu, or "wilde beeste" of the Hottentots

(*Connochetes Gorgon*); with horns 18 to 20 inches long, bent outward, the points bend over at acute angles towards each other. Of late years large quantities of the skins have come into commerce, as they make excellent bands for machinery. In 1871 as many as 34,622 dry gnu hides were sold in London. In 1874 no fewer than 20,000 gnus were destroyed in South Africa for their hides.

The GIRAFFE (*Camelopardalis Giraffa*) is hunted in Kordofan chiefly for its flesh, which is eaten—that of the young is said to be delicate—and for the stout skin, from which bucklers and sandals are made. The bones have also been imported for cutlers' use, as handles. The giraffes brought to Europe come generally from Nubia and Sennaar. In South Africa the giraffe has long since retired before the tide of colonial emigration, and is not to be met with south of Kolobeng, a point 380 miles north of the Orange River. The Hottentots in Southern Africa used to hunt the animal principally on account of its marrow, which is a delicacy they set a high value on.

The following list of other species of antelopes, with their common and scientific names, found in South Africa, may be found useful for reference, although they do not demand special notice or description:—Koodoo (*Strepsiceros capensis*, Harris), sable antelope (*Ægocerus niger*), roan antelope (*Ægocerus equina*), waterbok (*Ægocerus ellipsyprymnus*). Hartebeeste (*Acronotus* [*Alcelaphus*] *caama*), bastard hartebeeste or sayssabe (*Acronotus lunata*, or *Damalis lunatus*), pallah or red bushbok (*Antilope* [*Ægoceros*] *melampus*), bontebok (*Gazella* [*Damalis*] *pygarga*), blesbok (*Gazella* [*Damalis*] *albifrons*), bushbok (*Tragelaphus sylvaticus*), rheebok (*Redunca capreolus*), rietbok (*Redunca eleotragus*, *Eleotragus arundinaceus*, Gray), small rietbok (*Redunca isabellina*), rooi rheebok (*Redunca Lalandii*), oribe (*Antilope scoparia*, *Scopophorus ourebi*), duiker bok (*Cephalophus mergens*), steinbok (*Tragulus rupestris*), klipspringer (*Oreotragus saltatrix*, Harris), grysbok (*Calotragus melanotis*, Gray), bluebok or kleinbok (*Cephalophus caerulea*), and gemsbok (*Oryx Gazella*). The flesh of the gemsbok ranks next to the eland, and at certain seasons of the year they carry a great quantity of fat. Some smaller doubtful species have been described.

WATERBOK (*Antilope* [*Kobus*] *ellipsyprymna*, Ogilby).

CHAPTER VI.

FURS AND THE FUR TRADE.

In this chapter the early use of furs is glanced at, the great Arctic hunting fields of Europe, Asia, and America, are described—the trade and statistics of furs given in detail—the dressing, dyeing, and preparing of skins for furs. The trade carried on by the Hudson's Bay Company and dealers in the United States, Russia, and Germany, receives notice, with the value of imported furs, preparatory to detailed accounts of the principal fur bearing animals, to be described in subsequent pages.

ALTHOUGH nearly all the orders of the Mammalia supply peltries, it is those of the Carnivora and Rodentia which are chiefly valuable to commerce. Peltries is the name given to skins prepared with the hair on, intended for furs.

To trace the origin of the trade in skins and furs would imply a study on the origin of the human race. Necessity, the mother of invention, soon suggested to the inhabitants of the globe that as nature had not clothed them with sufficient warmth they had better appropriate the skins of animals well provided in this respect. Unaided by experience, without defensive or offensive arms, possessing no knowledge of the different metals which modern society has converted into fearful weapons of destruction, pursued by the large animals, reduced to inhabiting crevices in the rocks and the borders of lakes and rivers for the sake of shelter—man, by his physical organization, was obliged to declare war against all the beings of creation, attacking some as useful auxiliaries to satisfy his daily wants, and others to help him to wage implacable and terrible war against those animals whose nature and instincts precluded them from participating and assisting in the organization of society. The victims of the war thus pursued, supplied him with his daily food, and their skins, detached from the flesh by means of sharpened stones, were dried in the sun and then energetically rubbed with the oil and grease extracted from the intestines of the slaughtered animals. With porous stones a polish could be added to the skins, and thus the hides of the bullocks, horses, stags and other large animals were preserved, and employed to make the tents which sheltered the early patriarchs. Lions, tigers, panthers, the whole feline race, together with smaller animals, and even birds, helped either to shelter or clothe man, and afforded wearing apparel in which he was able to glorify himself in the presence of his fellow men.

Gradually the skins were freed from the adhering particles of flesh which invariably remain when the skins are merely wrenched off the dead animals. With bone, stone, and iron instruments this inconvenience was obviated, and then the skins were washed in water so as to rid them of the blood and open the pores and cleanse them of the dust and dirt incrusted thereon, and afterwards exposed to the frost. Having achieved this much, it was discovered that the skins could be greatly improved by plunging them in water containing a solution of alum and then putting them in vinegar. These baths prevented the skins from rotting, then they were dried in the shade and moistened again, and beaten and stretched, so that by dint of belabouring them, they were rendered quite supple, clean, and free from that disagreeable odour appertaining to skins less carefully prepared.

As man became civilised he invented other kinds ot clothing, but as a subsidiary covering furs have always been a fashion, and formed an article of traffic.

The fur trade has made great extension and progress since the sixteenth century. In 1535 the French took possession of Canada and in 1553 the Russians established their first stations in the north. The taste for furs was developed by these new hunting establishments which arose in different quarters. At the same time as the French formed factories in Canada and on the Mississippi, the British commenced a large trade in Hudson's Bay. In the eighteenth century the North West Company was established and commenced a powerful competition with that of Hudson's Bay. About the same time the Russian North American Company was formed, with its seat at New Archangel, to develop the fur trade of the countries from the Oural to Kamschatka, the north-west coasts of America, and the Aleutian islands, trading with the Chinese by Nijni Novgorod and Irkutzk, and exchanging furs for tea at Kiachta. In 1800 Astor the great trader started the American Company, and in 1825 the North-West Company was united with that of Hudson's Bay.

At present there remain only the Hudson's Bay Company, which explores and hunts over some three and a half millions of miles of British territory in North America; the American Company, which carries on the chase over that part of Arctic America, which the United States has bought from the Russians; and the Greenland Company, whose seat is at Copenhagen, and which prosecutes the fur trade in Greenland, Iceland, and the Polar sea. What has not the fur trade done for commerce, in pushing the explorations of the fur hunters, and improving our knowledge of geography, natural history, and science, and what influence have not these discoveries had on the civilization of the world!

The fur trade includes three classes of dealers. The collectors of skins, the wholesale skin and fur dealers, and the furriers or makers up of furs. The price of furs necessarily varies greatly, from the simple rabbit-skin worth 4*d.* to the Siberian sable worth £20 or the black fox worth £40 or £50, and is also dependent upon supply and the variations of fashion.

A large proportion of the furs used in America and Europe, are cured and dressed here, for although many expensive furs are not used in this country, London is the great fur mart of the world.

The trade in made-up furs determines what kind of skin will be most needed; for fashion in furs as in other articles of ladies' dress is most fickle, and the favourite of last season may be the least sought after in this.

It is a remarkable feature of the fur trade, that almost every country or town which produces and exports furs, imports and consumes the furs of some other place, frequently the most distant. It is but seldom that an article is consumed in the country where it is produced, though that country may consume furs to a very great extent.

In the account of the Master of the Robes to Louis IX. of France in 1251, there is a charge for the fur lining of a surcoat of 346 ermines, for the sleeves and wristbands 60, and for the frock 336, or 742 ermines in all for a single dress.

The four noble furs of those ages were the sable, the ermine, the vair, and the gris. The three former of these represented the three fur colours admitted into their armorial bearings. Every one at all acquainted with heraldry knows that ermine is represented by a white ground with black, somewhat lengthened spots. These were intended to designate the black-tipped tails of the animals, the skins being sewn together either with the tails on, or the tails were first cut off and afterwards sewn in rows upon the skins, sometimes alone, sometimes with a little wad of black lamb-skin on each side of the tail.

The vair was a squirrel, obtained probably at that time, as it is at present, from the southern provinces of Russia. It has a white belly, and a blue or rather dove-coloured back, on which latter account its colour, when blazoned, was azure. When these skins, entire, or at least only reduced to square pieces (called in ancient heraldry, pannes) were sewn together, the result was a varied surface of bluish grey and white, in alternate, somewhat bell-shaped figures; but as the white of the squirrel's belly is far inferior to that of the ermine, it was the custom for the more sumptuous kind of garments, to use only the back of the squirrel, and to form the alternate white figures of ermine.

Respecting the "gris" heraldic antiquaries seem much in doubt.

Some suppose it to be only the blue or grey back of the same squirrel as has been just described by the name of vair. This, however, is not very probable, especially as the common French name of the North American grey squirrel is *petit gris*, although in size it is equal to the vair. Greise and graies occur in two lists of furs inserted in M. Chancellor's travels in Russia and Muscovia, 1544; and as graye is the old English name of a badger, the heraldic gris is probably the fur of this very animal. If not it may be the calabar, or grey squirrel of Russia.

The Governor of the Hudson's Bay Company in one of their annual reports, commenting upon the opinion that the fur-bearing animals are dying out, and must soon be extinct, remarks :

"We think that those who argue in this way must forget the many thousand square miles of land in and beyond our territory which can never be settled, and whence these skins are mainly brought. No doubt the skirts of regions formerly resorted to by the Hudson's Bay Company only are now visited by other traders, but this Company still possesses and with proper management will long possess the best means for obtaining increased returns from the vast regions of Mackenzie River, East Main and other lands, where nothing is lost by the absence of settlement, because nothing will grow there."

Everywhere it must be remembered that these animals, like cattle or human beings, are liable to periodical failures of food, or periodical inroads of disease. Experience shows that their abundance runs in cycles. The failure in one year of an insignificant class of animal may cause the decrease in the next year of a far more valuable beast which feeds on the former. The whole chain of animal life is more or less linked together, and the different species as they depend on each other, fall off or increase again according as the supply of food and the vigour of each class may be more or less abundant.

The statistics which we publish show that instead of decreasing, the quantity of furs and skins of wild animals is increasing yearly.

We drive back wild animals on some points, but the globe is yet so thinly peopled for its size, that man does not arrest the production of animal life; on the contrary, for the last century and a half the quantity of skins collected has increased. Agriculture itself increases the production of these animals by augmenting their food supplies, and their reproduction is therefore greater than their destruction by the hunters who wage a continual war against them.

A large trade is carried on in what may be termed artificial products, in which common and cheap furs are so prepared as to resemble the rarer and costlier articles. The piecing of furs is a distinct branch in itself. Many articles of fur regarded as entire by the buyer are made up of several pieces, the size of the natural fur not sufficing for the purpose. The skill with which piecing is done is somewhat marvellous, and great care and judgment are required to assort the parts. All the clippings and cuttings of furs have their uses, and pass into different hands for various purposes. A great trade is carried on with fur remnants and cuttings to Greece; they are there joined into cloths and rugs with incredible patience and perseverance, and they also form linings for the national dresses.

Great skill and care are required in dyeing and tinting the skin. Usually the greasiness of the fur is killed or removed by a bath of lime-water, but this is now, in many instances, objected to. The seal-skin, for instance, should be carefully prepared for unhairing, it is then picked and beamed carefully, the fur cleansed of all fatty matter by repeated washings. The dye stuff is then applied daily with a brush, and the skin spread out in a drying-room at a suitable temperature. This latter process is repeated from 15 to 30 days in succession, till the colour has become sufficiently deep and dark; finally the skins are cleansed by thorough washings in pure soft water; and the product is a dyed fur seal-skin, whose lustre is clear, rich, and permanent. A gloss is imparted to the hair and at the same time suppleness to the skin, by treatment in cylinders with sawdust and sand, or the former alone. Dyeing by immersion is more practised, and better suited for the hides of domestic than of wild animals.

The fur seal has annually increased in demand for ladies' mantles, and like all other goods of a costly nature, has called forth an imitation of less value. This has been effected by removing the upper hair from the skin of the musquash, leaving the finer portion, which, having passed through the hands of the fur dyer, forms an excellent substitute at half the price.

The art of imitating, altering, and improving furs is carried on with very great success, so that ferrets are passed off for martens, and bear skins are deepened in tint and appearance. By means of certain dyes, dog skins and white rabbit skins are *tigred*, grey rabbits are turned into genette, the panther skin is imitated, and in fact all sorts of skins spotted and altered in appearance.

The most choice marten is the Siberian sable, but the Russians have found out how to dye the red marten skin as dark as the beautiful natural black sable.

It was lately mentioned in a New York paper that a case of Russian furs, in a camphor-wood box about 3 feet long, containing 400 small skins, and bearing the Russian official seal, was valued at nearly £3,000. Some of these skins were worth 10 guineas, some 3 guineas, others 5 guineas, and some were of fabulous prices. They are generally sold at a profit of 30 to 33 per cent. As it takes some 16 or 18 skins to make a complete mantle, and adding the workmanship and profit, a very choice set of furs may easily reach a value of nearly £300.

The immense tract of country in North America, to the north of the United States, belonging to Great Britain, may be regarded as an immense hunting ground which supplies a great part of the furs used in Europe and America; the northern and western States and territories of the American Republic also furnish a considerable quantity.

Among the most interesting and really instructive features of

Victoria, in Vancouver Island, is the "fur room" of the Hudson's Bay Company, within the Fort. Here may be seen great bales of the most valuable, rare, and beautiful furs and skins of the monsters of the deep and lords of the forest. Here we may stroke the back of the bear without fear, and fondle the polecat without contamination. It is like visiting a great menagerie comprising all the wild animals of Oregon and Washington territories, Vancouver Island, and British Columbia. Some of the skins have come hundreds of miles—from the wildest and most inaccessible places known to man What a volume of daring adventures, feats requiring coolness and courage, might be told of the capture and slaying of the wild beasts who once filled the glossy furs before you, and the flash ot

ARCTIC FUR-BEARING ANIMALS.

whose eyes once made the stoutest hearts tremble! Hundreds of white hunters and trappers are constantly employed by the company to hunt and trap the beaver, otter, black, grey, and silver fox, lynx, marten, wolverine or glutton, mink, musk-rat, &c., besides at all the trading posts these skins are eagerly sought after and purchased from the Indians; and a steamer, the *Otter*, is entirely engaged in going from one Indian port to another of the islands to the north of Victoria, trading blankets, biscuit, and other articles of necessity and prime value to the savages for these skins and furs. In Victoria is the grand dépôt where all the skins of the north-west territory find their way, are cured, packed, and forwarded to London once a year, in a vessel belonging to the Company, which arrives out in the month of March of each year, direct from the old country. The skins are regularly packed in large wine casks by means of a screw, and it is wonderful the number that are pressed into a single cask. The value of some of these furs in Europe is very great, and the profits of the trade are large.

Furs constitute one of the most valuable exports from the province of British Columbia, and the following is a list of animals the furs of which are obtained there, with their value :—

	Each.		Each.
Panther . . .	$2·50	Lynx . . .	3·00
Wild cat . . .	0·75	Common otter .	5·00
Wolf . . .	2·50	Sea otter .	50·00 to $80·00
Red fox . . .	25·00	Squirrel . . .	0·12
Fisher . . .	5·00	Red deer (elk) .	0·15 per lb.
Mink . . .	2·00	Blacktailed deer .	0·15 ,, ,,
Marten . .	5·00 to $10·00	Ermine . . .	0·50
Racoon . .	0·75	Fur seal . . .	10·00
Beaver . .	1·00 per lb.	Mountain goat .	2·00
Black bear . . .	5·00 to $8·00	,, sheep .	3·00
Brown bear . .	7·50		
Wolverine . . .	1·00	AT CARIBOO.	
Siffleur (marmot) .	0·50	Silver fox . .	$50 to $70
Musk-rat . .	0·25	Black fox . . .	100·00

Until the year 1860 the fur trade was entirely monopolised by the Hudson's Bay Company; but since that date the trade has ceased to be exclusively in the hands of this Company, and there are now a large number of persons who have invested capital in it. In general the trade is carried on by coasting vessels, which exchange goods for peltry. It is rather difficult to arrive at a correct estimate of the value of the furs exported; but it was stated at $210,000 in 1868, and $233,688 in 1869.

The prices of skins are regulated by the quality of fur and also by the condition of the pelt; if they have been torn in the trap, or riddled by shot, or injured in handling, they cannot take rank among No. 1 skins, no matter how fine the quality of the fur. In this respect there has been a great change of late years, by the improvement in traps and the doing away with the "dead-fall" and kindred arrangements. Cased skins, that is those which are not cut open on the belly, command the best prices. Those preferred "cased" are mink, musk-rat, otter, fox, fisher, opossum, and skunk. Skins that are well-stretched and dried, command better prices than those of the same quality that have been handled carelessly, and the fur dealers at the present time complain of carelessness in this respect. The Indian method of loosening the skin from the flesh is a good one; it enables the skins to be removed from smaller animals much cleaner than any other plan. They puncture the skin in two or three places where no injury will be done by the cut and insert a quill; by blowing through this quill the air is forced between the flesh and skin, which can be stripped off without the use of a knife, and comparatively free from the flesh.

The Hudson's Bay Company receive their supplies from the north-west coast of America, the produce of their hunters and traders on the west side of the Rocky Mountains, about July. The sales take place in August or September. Those from the eastern side of the Rocky Mountains come to hand in November, and are sold in London during the winter and spring.

The following is an enumeration of the skins the Company received and sold in the years named at decennial intervals:—

	1856.	1866.	1875.
Badger	1,105	508	2,001
Bear	9,255	7,855	5,898
Beaver	74,482	150,170	100,721
Fisher	5,182	4,420	2,186
Fox, cross	1,951	2,935	1,961
,, kitt	3,370	3,910	2,699
,, red	7,371	12.739	7,644
,, silver	613	761	603
,, white	10,292	5,524	4,333
Lynx and cat	11,634	34,615	15,661
Marten	179,275	115,667	61,782
Mink	61,516	53,044	62,760
Musquash	258,791	312,301	503,948
Otter	13,740	14,720	9,825
Sea Otter	290	85	11
Porpoise, half skins	483	471	131
Rabbit	90,937	144,519	48,291
Racoon	1,798	5,105	1,632
Seal, fur	36	1,845	1,427
,, hair	5,263	14,899	3,743
Skunk	11,319	2,755	2,331
Wolf	7,576	6,611	1,608
Wolverin	1,142	705	1,052
Total	757,431	896,164	842,248

Besides a few deer skins, swan skins, ox hides, &c.

The value of the furs and skins exported from the United States is shown by the following figures :

1821	766,205 $	1851	977,762 $
1831	750,938 ,,	1861	878,466 ,,
1841	993,262 ,,	1871	1,590,193 ,,

And in the three years ending 1874 the average annual export of furs exceeded 3,500,000 dollars or £700,000.

The value of the foreign furs imported into the United States in 1856 was stated at nearly £400,000, and these were chiefly to be made up for ladies' use.

The value of the furs and skins exported from Canada was in 1850, £19,395 ; in 1860, £68,661 ; in 1870, £63,816 ; and of the furs, dressed and undressed, in the year ending June 1874, £102,229, showing a great increase on previous years.

The prices of furs in Canada vary in different years. The following are the extreme rates in usual seasons for ordinary sizes and qualities—

Bear Skins . .	20*s.* to 80*s.*	Otter	25*s.* to 50*s.*
Lynx . .	12*s.* ,, 20*s.*	Mink	5*s.* ,, 10*s.*
Red Fox . .	5*s.* ,, 7*s.*	Elk & Moose, dressed	20*s.* ,, 40*s.*
Silver Fox .	50*s.* ,, 150*s.*	Hair Seal . . .	2*s.* 6*d.* to 5*s.*
Black Fox. .	150*s.* ,, 600*s.*	Stone Marten of Europe	20*s.* ,, 50*s.*
Beaver . .	3*s.* ,, 8*s.* per lb.	Red Marten ditto .	10*s.* ,, 20*s.*

The great centres of the fur trade in North America are New York, Boston, Montreal, and St. Louis; the furs are shipped to those cities from all parts of the fur-producing region, excepting the section controlled by the Hudson's Bay Company, and from these cities they are distributed to the various sections of this country and Europe. The principal furs obtained on the American continent are sable, mink, musk-rat, foxes of various kinds, racoons, fisher, skunk, opossum, bear, lynx, wolf, wolverine, buffalo, badger, beaver, otter, seal, rabbit, and monkey, chinchilla, coypu. Of sables America produces about one half the supply; of mink, four-fifths; musk-rat, more than nine-tenths; of chinchilla and coypu, all; about one-half the foxes, all of the racoon, fisher, skunk, buffalo, one-half the seal and from one-half to three-fourths of the other kinds. London is the great centre of the fur trade in racoon, beaver, mink, seal, fox, otter, musk-rat and wolves, although nearly all kinds of furs are found at the trade sales. Leipsic is a large market for the majority of furs obtained in America, and various kinds from Middle Europe, Russia, and Asia. Martens (fitch, stone and black), together with sable and ermine, are more largely dealt in than at other markets, while, as in London, almost all other kinds can be found. The rich furs of Russia are collected at Irbit, Siberia, and Nishni Novgorod, on the Volga River, where they are sold at the annual fairs. According to estimates made by experts, the total catch of a single season is about 20,000,000 skins; of these one-third are obtained in Northern

Asia and North-West America, nine and a half millions in North and South America, and three and three quarter millions in Russia, Sweden, and adjoining countries. The total value of these skins is not far from £2,600,000.

Among furs and skins used by the furrier are included those of the swan, penguin and grebe, and lamb skins, such as those of the white and grey lamb, the Persian lamb, and the same dyed. Of all these specimens will be found in the Collection of Animal products in the Bethnal Green Museum.

France furnishes from its own territory furs and peltries of the value of over one million sterling, and imports about double that amount; one half of the quantity is used up locally, and the rest exported. Those produced locally consist of about 2,000 martens, 36,000 polecats, 4,000 otters, 100,000 weasels, 60,000 foxes, 30,000 cats, 60 millions of rabbits, 72,000 goose-skins, 100,000 lamb, 40,000 sheep, and 25,000 goats', besides a quantity of hares, bears, wolves, ermines, badgers, minks, kids, &c.

It is in the governments of Archangel, Vologda, Viatka, and Perm, as well as in Nova Zembla, that the trade of the fur hunter is chiefly carried on in European Russia, to the value of about £64,000. In that part of the empire the bear, wolf, badger, fox, squirrel, and other animals are taken—the skins of which fetch various prices. The produce of the chase there, however, diminishes yearly. It is Siberia and the Russian company's possessions that furnish the larger part of the furs to commerce. In Siberia certain tribes pay their taxes in furs, and as these constitute the private revenue of the Emperor, the handsomest furs by consequence are never seen in the market.

Throughout the whole country furs are an object of pursuit; sables, martens, stoats, foxes, squirrels, and ermines, are tracked and trapped by hunters. As a general rule the furs of the eastern are of better quality than those of the western provinces, but the ermines near the rivers Irkutzk, Oby, and Ishim form an exception, being of three times the value of those found beyond the river

Lena. Still the time has not yet come for any suffering among the inhabitants in consequence of a scarcity of the animals. A large export goes on to the west; they are disposed of in Yakoutsk to the value annually of three or four hundred thousand pounds, and furs of more than twice that value are sold in a single town bordering on Chinese Tartary.

It is for these furs that the Chinese barter their own produce, and create the trade of second importance to Siberia.

The export trade of Russia to China was originally almost entirely confined to furs. The large quantity of Siberian furs paid to the Crown in form of tribute, and other supplies of the same article found an easy sale, not only in the interior, but in the markets of Central Europe, as well as in Persia and Turkey. But when the national costume was changed in Russia, and was replaced by that worn in the west of Europe, the supply became too large for the markets of Persia and Turkey, and the government, being an interested party, looked out for another market for its furs, which it found in the northern provinces of Mongolia and China.

Besides this tribute, Siberia, however, furnishes to general commerce ermine, marten, zibeline, petitgris, and fox skins, to the value of about £32,000. The skins of the wolf, fox, marten, and sea otter, are the principal, but these are diminishing year by year. Formerly they used to obtain about 40,000 skins of different kinds yearly, now scarcely half these are obtained.

To show the former extent of the trade in peltries or furs from Russia to China, it may be stated that the exports in 1837 consisted of the following number of skins :

Beaver	10,119
Squirrel	2,931,345
Otter, Russian	21,959
,, foreign	1,695
Sea bear	12,117
Steppe and other fox	134,671
Foreign fox	126,598
White fox	56,373
Fox paws or pads	529,575
Ukraine lamb skins	909,752
Trans-Baikal ditto	216,167
Foreign ditto	97,505
	5,074,878

The hunting and trapping for furs give large employment to the poorer classes of Siberia, and bring in an annual value of about £160,000. There used to be exported to China, via Kiachta, from one to three millions squirrel skins, 250,000 to 300,000 cat skins, and 260,000 fox skins, besides some beaver and lynx. The beaver skins have dropped from 10,000 or 12,000 to 3,000, the squirrel skins have declined altogether. The otter skins have fallen from 20,000 to 4,000 and 5,000, the bear skins from 12,000 to 5 or 6,000, the fox skins to 15,000, the cat skins have fallen to 8 or 9,000, the lamb skins from 1,250,000 to 74,000; and the lynx from 15,000 to about 2,000.

The following table was given in the French Jury Reports of the Paris International Exhibition of 1867, as an approximate estimate of the furs received in Europe from North America, Russia, and the Greenland Company.

Furs.	1800.	1825.	1852.	1866.
Beaver	164,237	76,277	51,280	154,971
Bear.	25,105	12,145	12,478	11,048
Otter	21,694	12,235	13,825	17,053
Pecar	6,598	7,644	9,449	6,157
Marten	70,053	100,574	116,804	124,484
Wolf	8,093	1,462	8,765	6,610
Glutton . . .	1,495	806	1,508	806
Lynx	19,708	12,252	12,710	38,751
Vison	9,344	46,176	209,125	60,169
Fox . . .	24,164	25,861	77,055	64,247
Marmot . .	109,979	52,721	562,177	388,178
Musquash . .	27,272	281,416	1,637,340	1,577,707
Deer and elk . .	29,288	66,351	54,406	31,812
Bay Lynx . . .	—	—	6,125	4,752
Opossum . . .	—	—	14,444	218,144
Skunk	—	—	1,618	74,591
Rabbit. . . .	21,825	—	54,827	144,519
Badger	—	—	1,710	618
Seal	—	—	17,151	34,775

Value of the imports into United Kingdom of furs of all sorts.

1870. . .	£659,872	1874	£617,276
1871 .	541,209	Articles made of fur . .	74,263
1872.	714,967	1875	971,697
1873	419,104	Articles made of fur . .	122,376

CHAPTER VII.

CARNIVOROUS MAMMALS.—(*Continued*).

The general history and statistics of the fur trade having been given in the preceding chapter, detailed accounts are now furnished of the principal carnivorous animals and their economic relations to man—commencing with the felines or cat tribe, including the lion, puma, panther, leopard, jaguar, ocelot, wild and domestic cats, chetah, European lynx, American Bay lynx, Canadian lynx, and Civet cats.

Carnivora.—The distinguishing characteristics of the Carnivorous Mammals, it has been well observed, are found in the perfection of structure-arrangement, number, and development of the teeth, the canine teeth being especially adapted for destroying other animals, and tearing, crushing, and dividing the flesh upon which these animals subsist. In the bears and their allies, which exist on a more or less mixed diet, the crowns of the molar teeth are furnished with small tubercles or prominences, evincing an adaptation to a vegetable diet.

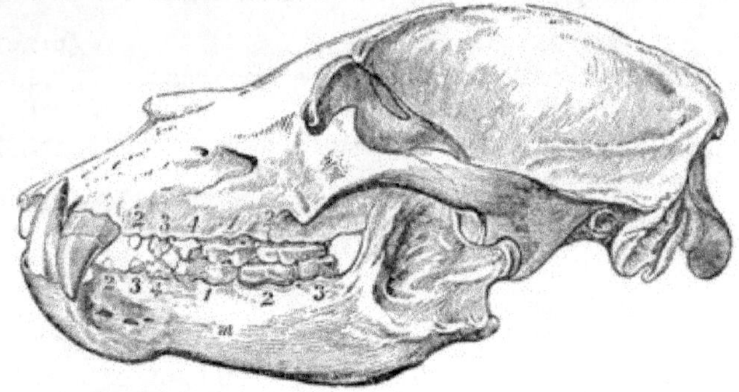

SKULL OF BEAR (*Ursus*), SHOWING THE DENTITION.

The inquiry may possibly be made, What are the uses of the numerous carnivorous animals which roam over large tracts of country, and are a terror to man and beast? But it has been often shown that the races of herbivorous animals, without a natural check from predatorial enemies or other slaughter, would soon become too numerous for the substances which have been allotted for their nourishment, and, by creating famine, would be the cause of their own destruction.

In their uses to man, this balancing of creation, as it were, is most important, by keeping down many that would desolate by their ravages pastoral regions. The more direct benefit will be found in the extensive commerce which is maintained with their skins, which form comfortable protections from cold and the inclemencies of the weather, and are beautiful and ornamental articles of winter dress.

There is a wholesale and retail trade in wild animals, and agents are at work for the wild beast dealers in every quarter of the globe. Travellers are dispatched to pick up strange animals in Central Africa, the Indian Archipelago, or South America, just as other traders send their buyers to Paris or London. They have dealings with the various Governments and Zoological Gardens of Europe. Zebras will be sold at £450 to £500 the pair; gnus for £170; rhinoceroses at £1,200 the pair; tigers at £300 each.

THE FELINES OR CAT TRIBE.—First in the list of the Felines stands the LION, of which there are two marked species, *Leo Africanus*, and *Leo Asiaticus*, inhabiting the greater part of Africa, and the warmer districts of India. There appear to be several varieties, if not species, of the African lion, as *Leo Barbarus* and *L. Gambianus*.

It is principally for its skin that the lion is sought, although living animals are valuable for menageries and zoological collections. In some years 100 to 200 skins are secured. The flesh of the lion is eaten by the Hottentots; and a tribe of Arabs between Tunis and Algeria, according to Blumenbach, live almost entirely upon it when they can get it. When a lion has been killed and the skin removed, the flesh is divided, and the mothers take each

NATIVES KILLING A LION AT MIDNIGHT, IN AFRICA.

a small piece of the animal's heart, and give it their male children to eat in order to render them strong and courageous.

They take away as much as possible of the mane, in order to make armlets of it, which are supposed to have the same effect.

It would seem from the Journal of the Marquess of Hastings, that this superstition as to eating lion's flesh is as strong in India. On the death of a lion it is stated—"Anxious interest was made with our servants for a bit of the flesh, though it should be the size of a hazel-nut. Every native in the camp, male or female, who was fortunate enough to get a morsel, dressed it and ate it. They have a thorough conviction that the eating a piece of lion's flesh strengthens the constitution incalculably, and is a preservative against many particular distempers. This superstition does not apply to tiger's flesh, though the whiskers and claws of that animal are considered as very potent for bewitching people." But the flesh of the lion has also been eaten with gusto by Europeans, for Madame Bedichon in her work on Algeria, states, that at Oran a lion was killed which three days before had eaten a man, and the Prefect gave a grand dinner, the principal dish being the lion, which the French gentlemen assembled ate with the greatest relish.

More recently still,—within the last year or two,—a magnificent quarter of lion, shot in the neighbourhood of Philippeville, Algeria, by M. Constant Cheret, was sent to the Restaurant Magny, Paris, and served up to a party of nineteen guests, who enjoyed with gusto "Estouffade de lion à la Méridionale" and "Cœur de lion à la Castellane."

Puma or Couguar (*Felis concolor*, Lin).—This animal, sometimes called the South American lion, is most common in the southern part of the continent, although its geographical range is said to extend to the north. The skins are chiefly used for carriage wrappers. The fur is thick, close, and of a reddish brown colour, approaching nearly to the colour of a fox on the back, and changing on the belly to a pale ash. When at a mature age, however, its general colour s a silvery fawn. Three or four other species are said to inhabit Paraguay, Buenos Ayres, and Chili.

The puma is very destructive to sheep, and has been known to kill fifty in a night, merely to suck a portion of their blood. The length of the adult animal is a little over four feet, and its tail two, to two and a half feet. The flesh is said to be tender and well flavoured.

TIGER (*Felis Tigris*).—This animal is exclusively confined to the Asiatic continent, and the islands of the Eastern Archipelago. The peninsula of Hindostan seems to be its great nursing place, but it extends far up into the hill regions. The district of Goalparah in Assam is so infested by wild animals, that £1700 has been paid in a single season for the heads of beasts of prey. And yet the reward for tigers is only ten shillings per head. The *Delhi Gazette* states that 250 tigers' heads have been brought in by natives in a single month. They are mostly killed by poisoned arrows from spring bows fixed near their favourite haunts. In 1871, 1100 wild beasts were destroyed in the Madras Presidency, at a cost of £2,200, of which 666 were cheetahs, 205 tigers, 129 hyenas, 97 bears, 12 wolves, and a few jackals. On the Nielgherry Hills 16 tigers and 29 cheetahs were killed, and £110 paid in rewards. In Kistna and South Canara £73 were paid for killing 146 tigers and leopards, but not before they had destroyed over 1760 head of cattle.

In China the mandarins cover the seat of justice with the tiger skin, and the easterns are very proud of their tiger skin rugs. A good skin is worth £3 or £4, but varies according to size and condition. They are sometimes obtained exceeding 11½ feet long. There are three good stuffed heads of the Bengal tiger (*Tigris regalis*) in the Bethnal Green Museum.

PANTHER (*Felis Pardas*).—This animal, principally found in Africa, is believed to be only a variety of the leopard, but differs in its superior size and deeper colour from it. Its colour is of a bright tawny yellow, with rounded black spots disposed in circles about his body; the breast and belly are white.

In Algeria recently a gun of honour was presented to a great slayer of panthers, Si-el Moufok, aged 40, who had killed 42 of these wild beasts, and his father 75 before his death in 1850. Si-el Moufok has a young brother who has already killed 3 panthers.

LEOPARD (*Felis Leopardus*, Cuv., *Leopardus varius*).—This animal, which is about four feet long, is dispersed over Africa, Asia, and some of the Indian islands. There are one or two varieties. Their skins, which are very valuable, are of a bright tawny hue, marked with black spots. There is a leopard skin shown with the furs in the Museum Collection. In this country the collocation of the leopard under the officer's saddle is a distinguishing mark adopted by some of the cavalry regiments.

The imports of these skins is but small in numbers, seldom averaging more than 100 a year, although in some years 150 have been imported. They are sometimes used on draught and saddle horses, and for ornamenting caps, fur garments, and cuffs. Leopard skins command good prices for hearth-rugs and military purposes, and for the seats of some carriages. The skin is worn as a mantle by the Hungarian nobles who form the royal body-guard of Austria.

In the Museum Collection is a fine African leopard skin rug bordered with bear skin, and four or five very choice Bechuana mantles or karosses made of leopard, and various other skins of native animals.

There is another species of leopard known as the ounce (*Leopardis uncia*), met with about the shores of the Persian Gulf, the fur of which is rather paler, rougher, and thicker.

JAGUAR (*Felis* [*Leopardus*] *onca*), or American Panther. This animal is as large as a wolf, and is formidable for its strength and ferocity, in these points resembling the royal tiger of Bengal. It has a fur, the ground colour of which is a pale brown yellow, very beautifully marked with chocolate brown spots and with streaks and

FIGHT WITH A LEOPARD.

stripes. It is used for ornamental purposes, as hearth-rugs, &c. It seems to be a merciful dispensation of Nature that the most terrible quadrupeds are not gregarious, but hunt alone or in couples. If lions, tigers, and jaguars herded like wolves, whole provinces would be depopulated by their ravages, and man would hardly be able to hold them in any subjection. But by destroying them in detail, their numbers can be kept within bounds, and their depredations confined to their native forests and jungles.

An oil from the adipose tissue of the jaguar is used externally in Brazil in cases of rheumatism, and also for fomenting boils. The Gauchos or herdsmen differ in their opinion whether the jaguar is good eating, but are unanimous in saying that cat is excellent.

Mr. Wallace, when travelling up the Amazon, one day had some steaks of the jaguar on the table, and found the meat very white, and without any bad taste. "It appears evident to me," he adds, "that the common idea of the food of an animal determining the quality of its meat is quite erroneous. Domestic poultry and pigs are the most unclean animals in their food, yet their flesh is highly esteemed, while field rats and squirrels, which eat only vegetable food, are in general disrepute."

OCELOT (*Felis Pardalis*).—This animal, rather less than the ounce, is a native of tropical South America. Its skin is beautifully variegated, of a bright reddish colour, with stripes of a deeper tinge edged with black variously disposed over the upper parts of the body, and hence is in great request by furriers. Two other species, the linked and the long-tailed, are enumerated as belonging to South America.

The skin of the bush cat (*Leopardus serval*), the tiger cat of the furriers, is also in some request, being of a golden grey, with dark spots and stripes.

There are some other species of Felines described by naturalists, such as the Margay (*F. tigrina*), and two or three species of Indian cats; but as these have no commercial or economic importance, it is unnecessary to describe them.

The WILD CAT (*Felis Catus*), from which the well-known domestic animal is descended, still inhabits the mountainous parts of Britain and Northern Europe. Its fur is much longer and rougher than that of the tame cat, grey, mottled and spotted with black, and its softness and durability render it very suitable for furs.

THE COMMON CAT (*Felis domesticus*, Lin.).—Besides its uses as a household pet and for keeping down rats and mice, there is a large trade carried on in the skins of cats for their fur. Cats have even been shipped in large numbers to Australia, California, Malta, and other places where rats have become too numerous to be pleasant. In some countries, as in France, a tax is levied on cats. In that country about 30,000 skins are furnished annually to commerce.

As there is scarcely a house in which one or more cats are not kept, and very many are also maintained in stores, warehouses, docks, &c., to clear mice, there cannot be less than four to five millions domestic cats in the United Kingdom, so that there must necessarily be a good supply of skins.

The common cat is fed on fish, and bred for its fur in Holland, where the finest skins are obtained. Large quantities are also collected in Holstein, Bavaria, Switzerland, &c. This fur is now greatly valued, and the supply of good skins continues far short of the demand. The black, spotted, and striped varieties are all much in request, to be made into wrappers for open carriages, sleigh coverings, and railway travelling.

In 1856 we imported 13,451 cat-skins, chiefly from the Continent, in 1856 32,138, and in 1860 9,741.

According to the "Bulletin of the Society of Acclimatisation" of Paris, the domestic cat is more eaten as food than is generally supposed. In Williams's "China" it is mentioned that wild cats are sometimes caught, and are considered a great food dainty.

THE CHETAH OR HUNTING LEOPARD (*Gueparda jubata*; *Cynailurus jubatus*, Wagler).—This animal is found in all the warm parts of the old world, from the Cape Colony to Persia and India. In Asia it is trained for hunting antelopes. The skin is of a light yellowish fawn colour above and white below, with the back and sides covered with black spots. It is used for ornamental purposes.

LYNXES.—There are several species of Caracals or Lynxes

EUROPEAN LYNX (*Felis lynx*).

inhabiting Africa and Asia, but the European and Canadian ones are those chiefly hunted for their furs.

The skins of the common lynx and the spotted lynx are shown among the furs in the Museum.

The European lynx (*Felis lynx; Lyncus virgatus*) ranges over a good part of Europe and some of the northern parts of Asia.

The length of the body and head is about three feet. The fur of its winter coat is in demand for various purposes. It is of a darkish grey, tinged with red, with dark spots and patches. There is a species which extends more to the south, known as *Lyncus pardinus*.

The AMERICAN BAY LYNX OR WILD CAT (*Lynx rufus, Felis Canadensis*, Geof.) differs from the common lynx in having shorter fur and longer pencils to the ears. There is a variety of this, perhaps even a distinct species (*L. maculatus*), whose fur, spotted with brown, is as valuable as the others. The specimens of the Bay lynx from the Columbia River are generally carried direct to China, without passing through the hands of European furriers. As a rule, the colder the climate the fuller and more valuable the fur. The lynx is found on the Mackenzie River as far north as 66°, and is not uncommon in the woods of Canada, especially in the Lake district.

The Canadian Lynx is said by Temminck to be identical with the lynx of Northern Europe. It has a light, though warm, hoary fur; the natural colour is of a beautiful grey or rusty brown, spotted with dark and rufous. Dyed of various colours, it is much used by the Chinese, Greeks, Persians, and others, for cloak linings, robes, and muffs, &c., being exceedingly soft, warm, and light.

The imports of these skins by the Hudson's Bay Company range between 5,000 and 16,000 yearly. In the 13 years ending 1868 the total imports by the Company had reached 330,000 skins. Of late years there has been a large increase in these skins. A few hundred are obtained annually in Alaska. In 1851 we imported 8,415; in 1858, 15,688; in 1861, 8,415; in 1875, 15,000.

CIVET CATS.—The AFRICAN CIVET (*Viverra civetta*) is most abundant in the hottest parts of Africa and in Abyssinia. It is chiefly remarkable for the highly odoriferous secretion, from which the perfumers used to prepare the old-fashioned "musk." Musk has now given way to some extent to patchouli, and to far purer and more delicate floral perfumes; so that civets are no longer in the demand which former periods of fashion created for them.

The skin of the civet has a yellowish-grey ground colour, with long hair on the back, large dusky spots disposed in longitudinal rows on each side. Furriers often confound this skin, which is light and soft, with that of the Zibet and Genet.

The pouch situated near the genitals is a deep bag, sometimes divided into two cavities, whence a thick, oily, and strongly musk-like fluid is poured out. When good, this odoriferous substance

ANIMALS FURNISHING CIVET AND MUSK.

is of a clear yellowish or brown colour, and of about the consistence of butter; undiluted, the smell is powerful and very offensive, but when largely diluted with oil or other ingredients, it becomes an agreeable perfume. Important medical virtues were formerly attributed to the civet; it, however, no longer forms an article in the materia medica, and even as a perfume has been laid aside.

Until the time of Buffon, the difference between the African and Asiatic species was unobserved, both being of nearly the same form and colour, but the number of dark marks on the tail is different in the two.

Zibet (*Viverra Zibetha*).—The skin of the Asiatic species differs somewhat in its markings from its African congener. The animal furnishes the same secretion, and has been found in the Philippine Islands. The caudal rings of this species hold an unvarying character, and are uniformly six in number, pale, upon a black ground.

The Muskars, a low class of woodmen, eat their flesh. In South India the secretion is much employed by the native practitioners under the name of kustre. In Travancore there was, and probably is still, an establishment kept up at the expense of the Government, in which these animals were kept and reared for the sake of their secretion. It has a disagreeable ammoniacal odour, and acrid, pungent taste. There is another species called the Rasse (*V. malaccensis*) found in Java, Singapore, and other Eastern districts, which yields the same perfume.

Genet or Spanish Cat (*Viverra genetta*, Lin.).—This animal, about the size of a small cat, has a beautiful soft fur of a pale reddish grey, the sides of the body being spotted with black, and a dark line running along the back. It is a native of the western parts of Asia, and is found also in Spain. The skin makes a soft

CIVET CAT.

and light fur, but it is imitated by the furriers with grey rabbit-skins dyed. The civet furnished by this species is less powerful than that of the others. There are one or two other varieties of genets.

The WILD CAT of Formosa (*Felis viverrina*, Hodgson) is much sought after by the Chinese for its soft pretty skin, to make cuffs and collars for their coats; *4s. 6d.* being the usual price for a single skin. In the collection of furs in the Museum the skin of the wild cat of India and of the black cat will be found.

The skin of the *Viverra pallida* is also valued for lining to great-coats. The poorer classes, who are unable to purchase the dearer furs, make use of these cheaper yet pretty skins. The Chinese eat the flesh of this animal, although it has a strong civet odour.

CHAPTER VIII.

CARNIVOROUS MAMMALS—*continued.*

This chapter deals with the dog tribe—the varieties of the domestic dog, the Esquimaux dog and dog sledges. The tax on dogs in various countries. The wolf—statistics of the ravages it commits, and the use made of its skin. Foxes—variety and value of the skins of this animal as furs, wholesale trade in them by the Hudson's Bay Company—statistics of imports. Racoon and bear skins. Badgers.

THE DOG (*Canis*).—All zoologists agree that there is no trace of the dog to be found in its primitive state of nature, although wild dogs exist in India, America, and Australia. Some have been led to believe the wolf is the original dog, but none of the native wild dogs have ever returned to the true form of the wolf. This question, however, is one not requiring to be discussed at any length here.

Sir John Richardson states that the great resemblance which the domestic dogs of the aboriginal tribes of America bear to the wolves of the same country, was remarked by the earliest settlers from Europe, and has induced some naturalists of much observation to consider them to be merely half-tamed wolves. The Esquimaux dogs are not only extremely like the grey wolves of the Arctic circle in form and colour, but they also nearly equal them in size.

If we consider the numerous varieties of the dog, from the King Charles to the Newfoundland ; the harrier and the bulldog ; those raised for food in China and the South Sea Islands to the sheep dog—from the smallest to the largest—the variation in size is fully one hundred fold.

There is more apparent difference between one of these dogs and the wolf and the fox, which are its allies, than between other animals far separated from the class. What modifications we find in the breed, what races lost, what races created? Like man, whom he follows over the globe, the dog changes like him, but more greatly, and readily adapts itself to all climates and all habits.

And why? Because long domesticated, it has experienced infinite changes according to the localities; because man has selected the differences in individual births to create varieties; because he has coupled dogs having the same instincts, and improved them so as to obtain a race suited for a special purpose,

which he has not ceased to improve. He has obtained from a primitive animal, which is no longer met with, a varied race of animals which nature did not form.

It will be sufficient for our present purpose to class dogs under three heads:—

1. Farm dogs, which includes the colly, the shepherd's and drover's dog, the mastiff, and the bulldog.

2. Hunting dogs, as the terrier, the hound, the harrier, the beagle, and the greyhound.

3. Shooting dogs, as the pointer, the setter, and the spaniel.

The dog is old at fifteen years, and seldom lives beyond twenty. Dogs are encouraged and kept in most countries as companions, housedogs, guardians for sheep, for the chase, and in many cases for beasts of draught.

In this country the employment of the dog for drawing burdens, or as a turnspit, has been done away with; but on the Continent they are still harnessed to small carts, and dog-trains for sledges are much employed in the Arctic Regions.

In Siberia, Greenland, and other northern countries, five dogs are yoked to a sledge, two and two, with the fifth in front as a leader. In general, only one person rides in a sledge, who sits

ARCTIC DOG SLEDGE.

sideways, and guides the animals by reins fastened to their collars. Such is their fleetness, that they have been known to perform a journey of 270 miles in three days and a half, and so great their strength, that they will convey a sledge containing three persons and their baggage sixty miles in a day over the snow.

THE ESQUIMAUX DOG (*Canis familiaris*, var. *borealis*).—Captain Lyon thus describes the Esquimaux dog: In form he is very similar to our shepherds' dogs in England, but is more muscular and broad-chested, owing to the constant and severe work to which he is brought up. His ears are pointed, and the aspect of the head is somewhat savage. In size a fine dog is about the height of the Newfoundland breed, but broad like a mastiff in every part except the nose. Young dogs are put into harness as soon as they can walk. Every dog is distinguished by a name, and the angry repetition of it has an effect as instantaneous as an application of the whip, which instrument is of an immense length, having a lash of from 18 to 24 feet, while the handle is one foot only. With this, by throwing it on one side or the other of the leader, and repeating certain words, the animals are guided or stopped.

ESQUIMAUX DOG.

I found (adds Captain Lyons) by several experiments that three of my dogs could draw me on a sledge weighing 100 lbs. at the rate of one mile in six minutes; and as a proof of the strength of a well-grown dog, my leader drew 196 lbs. singly, and to the same distance, in eight minutes. At another time seven of my dogs ran a mile in four minutes, drawing a heavy sledge full of men. Afterwards, in carrying stores to the "Fury," one mile distant, nine dogs drew 161 lbs. in the space of nine minutes. My sledge was on runners, neither shod nor iced; but had the runners been iced at least 40 lbs. might have been added for each dog.

The flesh of the dog is eaten in several countries, and its skin utilised for leather. The flesh of the North American dog, var. *Canis canadensis*, is esteemed before that of almost any other animals by the Canadian voyageurs, and is eaten by some of the Indian tribes on the Saskatchewan and shores of the Great Lakes; but the Chipewyan tribes hold the practice in abhorrence, because they consider themselves to be descendants of a dog.

In many countries, to keep dogs within due bounds, a tax is levied on them, as in France, the United Kingdom, and parts of Germany. The dog-tax in Great Britain is 5*s*. per annum, and it realized in the year ending December, 1874, £313,017, so that there was duty paid on 1,252,068. The Commissioners of Inland Revenue, however, complain that, notwithstanding considerable exertion on the part of their officers and the assistance of the police, there is still a large number of dogs uncharged.

There are more than 120 packs of hounds kept, the number of harriers is about 20,000, and other fancy breeds and dogs for the chase, 114,500. In Yorkshire there are ten packs of foxhounds, one pack of staghounds, and five or six of harriers, equal in all to thirteen or fourteen packs of foxhounds. Thirteen packs of foxhounds, or fifty couple each, viz. 1300 hounds, consume annually 200 tons of oatmeal, at a cost of £2600, besides the carcases of about 2000 dead horses.

In Ireland the dog-tax is only 2*s*., and in 1870 duty was paid on 270,422, being a dog for every twenty persons. The tax produced £27,042, and after deducting £7000 for expenses in the administration of the Act of Parliament, there remained £20,000, which was paid over to the authorities in reduction of local taxation. There are now about 300,000 returned in Ireland.

In France the number of dogs was estimated at 2,250,000. The tax in that country is about six francs a head. Many are thrown into the Seine by their owners rather than pay the tax, and these carcases are occasionally so numerous, and so fat, that at Javelle there is quite a trade carried on in fishing for and boiling them down for the fat, of which about 2,250,000 lbs. are obtained annually, this is sold to the glove-makers at 1s. the pound, for dressing the better class of straw-coloured gloves.

In the city of Berlin the dog-tax is three thalers, or about 9*s*., and it is found there are about 21,000 dogs kept; of these nearly 3,000 are used in drawing carts. It was thought that the imposition of the tax would check the number of dogs kept, but notwithstanding the tax and the obligatory use of a muzzle, they were found to have increased by one-third in four years. In Wurtemburg the dog-tax brings in 376,355 marks, or nearly £19,000; in the Hawaian Islands 20,000 dollars, or £4,000.

In the United States it is officially computed that there is a dog to each family, and this gives in round numbers 8,000,000 dogs, each of which costs eight dollars a year to keep, and consumes annually food sufficient to raise a pig. But it is believed there are twice that number in the States. The dogs are found to commit sad ravages among the sheep there.

And then it is reasonably argued what benefits might not result from killing the larger number off. For instance :—

6,500,000 dog-skins, worth . . .	$1,625,000
6,500,000 carcases, for chicken-feed . .	2,000,000
6,500,000 refuse for manure . . .	375,000
Total .	$4,000,000

Thus the surplus dogs, by proper use, might yield £800,000. If chopped up or ground, their meat and bones would be excellent for chicken feed; and their skins good for gloves and other manufactures.

The skins of big mastiffs are fit to be tanned for boots and shoes or thick riding gloves, the skins of lesser dogs can be dressed white for gloves. In the city of New York 6,000 to 8,000 dogs are annually impounded, a very small portion of which are redeemed. They are killed by drowning, and the carcases are taken to the offal boats, which convey them to Barren Island, where every part of them is turned to some useful account. The fat is rendered out, the skins are sold to glovers, and of the bones an excellent compost for fertilising land is made. There are about twenty-five regular dog-catchers, and a number of other persons who incidentally bring in those that they pick up.

In the Animal Products Collection there is a pair of child's cuffs made of the hair of a French poodle.

WOLF (*Canis occidentalis*, Baird).—The skin of the wolf is much valued, and is used as cloak and coat linings in Russia and other cold countries by those who cannot afford the more choice kinds; also for sleigh coverings and rugs for open travelling carriages and wherever additional warmth is required.

The varieties of the American wolf have finer fur than the European species; the Prairie wolf (*Canis lutrans*), supplies one of the furs of the Hudson's Bay Company. The wolf of the northern districts is covered with a long and comparatively fine fur, mixed with a large quantity of shorter woolly hair, and it has a more robust form than the European wolf. In some districts the wolves are very numerous, and vary greatly in the colour of their fur, some being white, others totally black, but the greater number are mixed grey and white, more or less tinged in parts with brown. According to a pamphlet which M. Lazarevsky has circulated, the wolves in 1873 did nearly as much damage as a Tartar invasion might have inflicted. They carried off 179,000 cattle and 562,000 smaller domestic animals from the forty-five governments of Russia in Europe. In the Baltic provinces 1,000 head of horned cattle were destroyed, and in the Polish provinces 2,700 oxen, and 8,600 sheep, pigs, and goats. If a cow be reckoned as worth thirty roubles, and a sheep at four roubles, the gross sum of the tribute levied by the wolves in Russia must reach 7,700,000 roubles (3*s.*). This is an amount of money quite well worth looking after, and it represents a number of wolves which must be dangerous even to human life.

Some singular details are given in a recent Russian paper of the ravages committed by these animals, which are worth reproducing. The number of wolves, it is said, in Russia, cannot be less than 170,000, and they eat of feathered game alone 200,000,000 head. In 1875 no less than two hundred human beings were destroyed by wolves. A comparison is instituted between the losses occasioned by cattle plagues and fires as against those caused by wolves, and, extraordinary as it may seem, the proportion of damage done by wolves as compared with cattle plagues is as 200 to 240. The amount that wolves will eat is enormous. In two or three

HUNGRY WOLVES.

hours a pair will eat the half of a horse weighing 700 lbs., they themselves weighing not more than 100 lbs.

In 1873, in one government, that of Vologda, wolves killed 14,000 head of large cattle and 35,000 head of small; in the Kazan government they killed 5,000 large and 26,000 small, of an aggregate value of £36,700. In the St. Petersburg government the losses are less, but even there, in the same year, property was destroyed by wolves to the extent of £10,000. In forty-five Russian governments, exclusive of the Baltic provinces and Poland, 714,000 head of cattle were destroyed in one year, making a loss to the country of over £1,000,000.

In the forests of France, and in the Pyrenees, the wolves occasionally attack shepherds, and they now and then venture within the walls of lonely chateaux and farm-houses. But their numbers of course cannot be compared with the enormous hosts of savage beasts in Russia, which one may perhaps guess at from the quantity of wolves that must band together to kill and carry off one able-bodied ox.

The European wolf has a coarser fur, with less of the soft wool intermixed with it, than the American wolf. Sir John Richardson enumerates the following varieties:—

a. Common grey wolf (*L. griseus*).	*d.* Dusky wolf (*L. nubilus*).
b. White wolf (*L. albus*).	*e.* Black wolf (*L. ater*).
c. Pied wolf (*L. sticte*).	*f.* Prairie wolf (*Canis lutrans*).

The skins of the American wolves are not split open like the large wolf-skins, but stripped off and inverted or cased like the skin of a fox or rabbit, and hence in the Hudson's Bay Company's lists called "cased wolves." From 4,000 to 12,000 wolf skins are imported annually by the Hudson's Bay Company. In 1851 there were but 2,400 skins received; in 1861, 3,669; in 1868, 9,193.

The hide of the wolf is considered peculiarly fitted for knapsacks and similar purposes, for which it is much employed in Germany. There is a wolf skin shown in the fur series at the Bethnal Green Museum.

The canine teeth of wolves are used, mounted in a handle, by gold workers in France, to polish and burnish gold. The jewellers and bookbinders especially use them, the latter for smoothing the edges of books gilded. They are also mounted for children to rub their gums with like coral to facilitate dentition.

Strings of the teeth of their enemies, and of bears, panthers and other wild animals are favourite ornaments on the necks of young men among the Dyaks of Borneo. In Case **168** in the Museum there is a necklace of bears' claws, and in Case **171** one of tigers' teeth, and another of monkeys' teeth. Even in civilised society tigers' claw-jewellery and the teeth of carnivorous animals are worn when handsomely mounted.

The Cayapas in the district of La Tola, Ecuador, male and female, wear round their necks large collars made of the teeth of the jaguar, caiman, snakes, &c., mixed with shells, eggs, &c.

FOXES.—The fur of the fox is in high esteem in Russia and China. The skin of the red fox (*Vulpes fulvus*, Baird) is finer than that of the *V. vulgaris* of Europe; its value is from two to three dollars; that of the cross fox (Var. *decussatus*), is valued at twelve to fifteen dollars; the white and black varieties are also highly prized according to the uniformity or intensity of the colour. The grey fox (*V. virginianus*) and the kitt fox (*V. volez*) have a much coarser fur. Fox fur is used much for sleigh robes, caps, and trimmings. There is a good series of fox skins in the Collection of Furs in the Museum.

The most valuable of the American fox skins is the silver grey, which twenty years ago fetched as high as eight guineas per skin; the black fox, which is rarely met with, formerly sold at £10 a skin; now they are worth four times that price; the cross fox at £1, and the red fox about £4. The Turks are the principal purchasers of the red fox skin. The wood grey fox skins, which are obtained in the States in considerable numbers, are much cheaper. This animal, so common about Buffalo and the south, is not known in any part of Canada.

A fine sinew thread taken from the tail of the fox is used by the North American Indians in their bead work.

The SILVER OR BLACK FOX (*Vulpes fulvus*, var. *argentatus*), when in first rate condition, produces one of the most valuable furs in the Hudson's Bay Company's territory, or that the world produces. A very limited number of skins only come into the market, 500 to 1,000 from America; black skins have been known to realise as much as £50 each, and even then are generally purchased for re-exportation, being highly prized in Russia. La Hontain speaks of a black fox skin as being, in his time, worth its weight in gold.

They are of an incomparable lustre and beauty. Some have been taken in Newfoundland. In its finest and best condition this animal is almost wholly black, excepting the tip of the tail, where the black fur is intermingled with white silvery hairs.

RED FOX (*Vulpes fulvus*).—This animal is indigenous to the wooded districts of the colder parts of North America, where it is found in immense numbers. The American fox has a much finer brush than the European species, and is a much larger animal. Its fur is of exceeding beauty, of a ferruginous colour. All the varieties of foxes are of one species, and live and breed together, but owe their difference of colour to unknown causes. From 8,000 to 30,000 are received yearly in the Hudson's Bay Company's ships.

CROSS FOX (*V. fulvus*, var. *decussatus*).—This variety is readily distinguished by a black cross on the neck and shoulders, from which it derives its name. It is probably a cross between the silver and red fox. This fox inhabits the northern parts of America. Its fur is a sort of grey resulting from an admixture of white and black hair. From 2,000 to 5,000 are obtained annually by the Hudson's Bay traders. In former years it was worth four or five guineas a skin, whilst that of the red fox did not bring more than fifteen shillings. The difference in value seems to depend principally on the colour, for some of the red foxes appear to have as long and as fine furs.

The KITT FOX (*Vulpes cinereo argentatus*) is the smallest of the American foxes. About 5,000 are received annually by the Hudson's Bay Company.

The ARCTIC FOX (*Vulpes lagopus*, Baird) exhibits in a remarkable degree that change of colour to which all polar animals are more or less liable. In winter it is pure white, in summer a line of a darker colour appears on the back with corresponding transverse stripes upon the shoulders, so that it is sometimes mistaken for the cross fox. The white skins are exported into the London market in considerable quantities. The fur is of small value in commerce when compared with that of any variety of the red fox.

Its flesh, on the other hand, particularly when young, is edible; whilst that of the red fox is rank and disagreeable. It is compared in flavour to that of the American hare, and resembles the flesh of a kid. About 5,000 skins are received on the average yearly by the Hudson's Bay Company, although in some years 12,000 have been sold. There is a darker variety known as the sooty or blue fox (*V. fuligonus*), of which very few are obtained. In Greenland, in the year ending March, 1875, 3,100 fox skins were obtained; of these the blue fox skins were to the value of £4,942, and the white fox skins £370.

The following Table shows the imports of the various kinds of fox skins by the Hudson's Bay Company for a series of years:—

	Silver.	Cross.	Red.	White.	Blue.	Kitt.
1856	613	1951	7371	10,292	103	3370
1857	1047	3188	10,484	4940	63	5776
1858	1060	3472	9706	2103	21	10,003
1859	1163	3982	11,489	1565	15	5547
1860	1177	4030	11,031	3355	43	4568
1861	1070	3403	8897	5084	42	2,532
1862	632	2248	7783	2805	23	2905
1863	587	1947	6402	3310	29	5532
1864	612	1963	5716	12,235	78	2409
1865	459	1800	8760	4821	33	3126
1866	569	1910	8125	5917	36	5141
1867	1089	4445	29,439	2113	—	5896
1868	1359	5132	22,065	10,341	—	6624

The Racoon (*Procyon lotor*, Cuv.) is found in the warmer parts of North America. The skins obtained from the western States, particularly of Michigan, are considered much better than those of Ohio, Pennsylvania, or New York. In 1840 they sold at a dollar a skin. Russia is the principal market, and during late years the price has fallen much lower. The skins afford a rather handsome fur of a greyish-red colour for robes, and were also employed in the manufacture of felt hats. They are used throughout Germany and in Russia as a lining to the long travelling coats and other equipments of northern countries.

The imports by the Hudson's Bay Company range between 1,800 and 4,700 skins annually; but large numbers also come in from the United States, bringing up the total to from 300,000 to 500,000 a year. This skin is shown in the Museum Collection.

RACOON (*Pro-yon lotor*).

BEAR-SKINS. The American black bear (*Ursus americanus*, Pallas) differs from the European species, and affords in season a thick and brilliant fur. Bear-skins are used as saddle-cloths for horses, for foot-muffs, and furs, grenadiers' hats, and formerly for cuffs. This fur takes dyes well.

Bears are found in considerable numbers in the Minnesota territory and other parts of the new settled States of America, also in small numbers in Canada and the Lower Provinces, but they are constantly diminishing before the progress of civilisation. The black bear is by far the most numerous, but few of the grizzly species being found. An average skin is worth five dollars—a very good one (she-bear), from six to seven. They are principally used for saddle housings and harness trimmings, and sometimes for sleigh robes. Their skins as furs are best when the animals are just issuing from their winter's sleep; and at that season the Indians are reaping their bear harvest.

About 8,000 bear-skins are annually imported by the Hudson's Bay Company, and some others are brought from the United States, Canada, &c.

In 1806, 9,255 bear-skins were received by the Hudson's Bay

Company; in 1866, 7,855. In 1870 we received from America 11,777 bear-skins, valued at £15,278. In later years the Board of Trade have not specified the separate skins imported as furs.

The BLACK BEAR inhabits every wooded district of the American continent, from the Atlantic to the Pacific, and from Carolina

A POLAR BEAR HUNT WITH ESQUIMAUX DOGS.

to the shores of the Arctic sea. The fur on the body is long, straight, shining, and black. The cinnamon bear is considered to be an accidental variety of this species. They hibernate in the northern fur countries for about six months, being in a fat condition; further south they only hide themselves for three or four months. Sir John Richardson tells us that at one time the skin of a black bear, with the fur in prime order and the claws appended, was worth from twenty to forty guineas, but at present the demand for them is so small, from their being little used either for muffs or hammer-cloths, that the best sell for less than 40*s*. There is a black bear-skin in the Collection of Furs in the Museum.

In the forty years ending with 1862, 10,000 black bear-skins were obtained in Alaska.

With the progress of settlement and civilisation, bears are fast passing away both in northern Europe and America. There was a time when as many as 500 bears were killed in one winter in two of the counties of Virginia. Then the Indians shared largely in the spoils as well as the excitement of the chase. Their mode of serving up the bear, as a first course, was to roast it whole, entrails, skin, and all, as they would barbecue a hog. Most of the American planters preferred bears' flesh to beef, veal, pork, or mutton. Bears' paws were long reckoned a great delicacy in Germany, and after being salted and smoked, were reserved for the tables of princes. The tongues and hams are still in repute.

The black bear is replaced on the barren grounds of North America by *Ursus Richardsonii*, a species bearing a strong resemblance to the *U. arctos* of Europe, another species the brown or barren-ground bear (*U. ferox*), and the grizzly bear (*U. horribilis*) ordinarily dwells among the Rocky Mountains. The latter is large, strong and ferocious. They are sometimes nine or ten feet long, and are reported to attain a weight exceeding 800 pounds. From the black, and indeed from all, the natives derive food; they also cut the summer hides into cords. The claws of grizzly and barren-ground bears are much prized for necklaces and coronets by the Indians, as a proof of their prowess.

The Polar Bear (*Ursus maritimus*, Linn.) is perhaps the only species common to both continents, and may be considered as a sea animal, inhabiting the ice floating between them. They come down on the ice drifted round Cape Farewell in the current from the east coast, and some are taken on the ice round Upernavick, in the far north.

In Capt. Hall's "Life with the Esquimaux," among other anecdotes given of the ingenuity of the polar bear, is one by which he kills a walrus when basking in the sun on the rocks. If this

POLAR BEAR KILLING WALRUS.

happens to be near the base of a cliff, the bear mounts the cliff and throws down upon the animal's head a large rock, calculating the distance and the curve with astonishing accuracy, and thus crushing the thick, bullet-proof skull. If the walrus is not instantly killed—simply stunned—the bear rushes down to it, seizes the rock, and hammers away at the head till the skull is broken.

The bear-skin is of the greatest value to the Arctic tribes. They dress it by pinning it down on the snow and leaving it to freeze, after which the fat is scraped off. It is then hung up to dry in the intense frost, and with a little scraping it becomes perfectly supple, both skin and hair being beautifully white. The Greenland Company get from forty to sixty white bear-skins annually.

The flesh is eaten by the Esquimaux and the Danes in Greenland, and when young, and cooked after the manner of beef steaks, is by no means to be despised, although rather insipid. The fat, however, ought to be avoided, as unpleasant to the palate. Dr. Scoresby tells us that the muscular fat of the bear is well flavoured and savoury. "I once," he adds, "treated my surgeon to a dinner of bear's ham, and he did not know, for above a month afterwards, but that it was beef steak. The liver is very unwholesome." Other arctic voyagers consider it exceedingly coarse. The Russian sailors who wintered at Spitzbergen found it to be much more agreeable to the taste than the flesh of the reindeer.

Sir John Richardson well observes, in his "Fauna Boreali-Americana," that when people have fed for a long time solely upon lean animal food, the desire for fat meat becomes so insatiable, that they can consume a large quantity of unmixed and even oily fat, without nausea. Our arctic seamen relish the paws of the bear, and the Esquimaux prefer its flesh at all times to that of the seal. Instances are recorded of the liver of the polar bear having poisoned people.

The BROWN BEAR (*Ursus arctos*), although at one time common in many parts of Europe, is now restricted to a few secluded valleys in the Alps, Pyrenees, and mountains of Norway and Lapland. The dark-coloured race, long considered a distinct species, under the name of the European black bear, together with the barren-ground bear of North America, are now included among the varieties of *Ursus arctos*. They are hunted for their skin and fat. The distribution of the brown bear is more extensive than any of the family. In Asia it inhabits Siberia and the Altai as far westward as Japan; when the Altai is crossed, and the great Himalayan chain examined, there is found another brown bear, which has been named the Isabella bear, from the prevailing light fulvous colour of the fur.

KILLING A POLAR BEAR.

As to the distinction between the *Ursus arctos* and the Isabella coloured bear of the Himalaya, the difference rests only in the colouring of the tips of the hair, and that is not always a sure criterion. No doubt the majority of the Isabella bears have the tips of the hairs brownish white, but brown is the fundamental colour. Some individuals, especially old males, are almost marone, whilst others vary from brown to brownish yellow, and are a dirty white; hence travellers speak of the "White and Brown bears of Cashmere": it is evident, however, that neither age nor sex determines the colour with any degree of accuracy.

The Isabella bear is found on the mountains of Armenia, where it has long passed under the name of the Syrian bear, and is perhaps the animal referred to in the Bible. It is not rare on the Caucasus and high ranges of Persia, Affghanistan, and Himalaya, at least as far eastwards as Nepaul and probably much farther.

The winter's coat is shed about midsummer, when the old hair and under wool, called *peshmena*, hang in matted masses on its sides. The bear's *peshmena* is not attained until autumn, and after the new coat has gained considerable length. It is analogous to that of the wild and tame goats and sheep of Thibet, and more or less pervades all the quadrupeds of the high and snowy ranges. The *peshmena* of the ibex is softer than that of the tame goat of Thibet. Its fur is thick and long in winter, but does not contain much under wool.

The Thibet bear (*Ursus tibetanus*) is a native of the lower Himalayan ranges, and is said to be found in Persia, Affghanistan and Northern China. There is a white mark on the chest shaped like the letter Y, the two legs proceeding a short distance up the side of the shoulders. Towards the end of October, after this bear has fed on fruits and grain, like its congener, it becomes very fat. The native hunters state that the kidney fat is useless as an article of commerce on account of being tainted with the smell of the animal's urine. They accordingly preserve only the external adipose on the loins and inside of the thighs.

THE EUROPEAN BADGER (*Meles Taxus*) differs totally from the American one in many points, especially in its dark-coloured, much coarser, and shorter fur.

The skin with the hair on being impervious to rain, used to be employed in France to cover trunks, the collars of draught horses, and their harness. The skins were also formerly made into pouches by the Highlanders. From the wiry nature of the hair it is extensively used for the manufacture of superior kinds of shaving brushes and artists' pencils. It is preferred in China to pigs' bristles. Brushes made of badgers' hair are used by painters in softening the colours employed in imitating wood. The flesh is reckoned a delicacy in Italy, France and China, and may be cured like hams and bacon.

THE AMERICAN BADGER (*Meles* [*Taxidia*] *Labradoria*, Sabine). The fur of this animal is very soft and fine, and about three and a

AMERICAN BADGER (*Meles Labradoria*).

half inches long on the back, of a purplish brown colour from the roots upwards, variegated with narrow black rings near its summit, and tipped with white, producing a pleasant and somewhat mottled or hoary grey colour, but exhibiting no brown tints when the fur lies smooth.

In the eleven years ending with 1866, the average annual import of these skins by the Hudson's Bay Company was from 1,000 to 1,700. The flesh is said to be not inferior to that of the bear, to which it is closely allied in structure.

CHAPTER IX.

CARNIVOROUS MAMMALS—*continued.*

The last chapter having nearly exhausted the description of the fur-bearing animals of commerce there remain but a few more to be spoken of; these include the Wolverine or Glutton, the Skunk, the Fitch or common Polecat, the Martens or Weasels, including the Ermine, Mink, Fisher, and Sable; the common Otter, American Otter, and Sea Otter; we then pass on to the varieties of Seal, and the description and statistics of this important fishery carried on for the skins and oil of the animals.

WOLVERINE, OR GLUTTON (*Gulo luscus*, Sabine).—This North American animal is nearly the size of our badger. It ranges from 75° to 42° N. It has two distinct kinds of hair, the inner fur being soft and about an inch long, the intermixed hairs are rigid, and about four inches long. Its fur is of a deep brown colour, passing in the depth of winter almost into black. The imports through the Hudson's Bay Company average about 1,000 to 1,500 skins a year. It is much used for muffs and sleigh robes; a small number are obtained in Alaska.

THE AMERICAN POLECAT (*Mephitis Americana*, Sabine), or SKUNK as it is invariably called in that country, has a long, soft, black fur. It can discharge, when molested, a fluid from a small bag, near the root of the tail, which emits one of the most powerful stenches in nature, that produces instant nausea. The odour has some resemblance to that of garlic, although much more disagreeable. Owing to the repugnant smell which the animal possesses, and which the fur, even when it has passed through the dressing process retains, it was long considered of small commercial value. Continued experiments, however, surmounted the difficulty; the two stripes of white coarse hair down the back are removed, and a skin of rich black fur is formed, so that it now takes rivalry with superior furs for ladies' wear.

The history of this skin is suggestive that there may be many other skins that could be rendered useful, were apparent objections sought to be overcome. Twenty years ago the number that appeared in the London market was trivial, about 2,000 to 4,000, in 1861, 11,000 were brought to public sale; in 1866, 11,319, were imported by the Hudson's Bay Company; and now that their preparation has been still further improved, this formerly neglected skin bears a higher value. When care is taken not to soil the carcase with any of the strong smelling fluid, the meat is considered by the natives to be excellent food.

The skins of the FITCH OR POLECAT OF EUROPE (*Putorius fœtida*) are used for making muffs and victorines. It has a fur of good quality, useful for common purposes. The long brown hairs which cover the inner yellow fur are used for making artists' brushes. A stuffed specimen of this animal is in the collection.

THE MARTEN.—The form of this animal is well known. It has a pleasing aspect. Its fur is about an inch and a quarter long, of a pale, dull, greyish brown, or hair brown colour, from the roots upwards, dull yellowish brown near the summit, and tipped with dark brown or black; the lustre of the fur is considerable. Being fine it is used for trimmings, and is also dyed so as to imitate sables and other expensive furs. Hence it has always been an important article of commerce.

The collective import of Marten skins from North America is always large, although it varies year by year. The quantity sold by the Hudson's Bay Company was, in 1851, 171,945; in 1856, 179,275; in 1857, 171,000. The next two years the average imports were 139,000, then in subsequent years they ranged between 74,000 and 112,000, and recovered to 125,000 in 1865, and 144,000, in 1866. In 1870, the number imported was only 79,674, valued at about £1 3*s.* each.

One of the Mustela tribe, the ferret (*Mustela furo*) is used for driving rabbits from their burrows.

The WEASELS of the fur countries become white in winter, like the ermine and are not distinguished from them by the traders. There is in the Museum Collection, a stuffed white weasel (*Mustela nivalis*) common to Russia and Sweden.

THE ERMINE OR STOAT (*Mustela* [*Putorius*] *erminea*) is a common inhabitant of America from its most northern limits to the United States. Brown in summer; in winter time the fur in some specimens is of a pure white colour throughout, except on the end of the tail, which is black. The skin is from eight to twelve inches in length. It so happens that royalty and nobility have adopted this fur in some countries, as one of their emblems, which has given to it a value far above its merits. It is made into various articles for ladies' wear, with the tail attached. Minever, the royal fur, is ermine with small pieces of black lamb inserted. The skins are called "clicks" by the Indians, and form a medium of currency. In 1870, 296,255 ermine skins were imported, valued at about 2*s.* each.

Among furs, that of the MINK (*Putorius vison, Vison lutreola*) stands pre-eminent for ornament, wear, and durability. It is of exceeding beauty, and increasing value, and the animal does not seem to diminish in numbers. The fur, of a chocolate or umber brown, is very fine, although short; it is of two sorts, a dense down, with longer and stronger hairs. Two varieties differing in size are found in Nova Scotia, the smaller, *nigrescens*, has the more beautiful fur. The *Putorius Cicognanii* and *Putorius Richardsonii* (Bon.), two distinct species, are also met with in that province; they differ from the true ermine chiefly in length of tail.

The Mink in 1840 sold in America as low as 6*d.* per skin; but in 1855 commanded as much as 7*s.* 6*d.* and was in active demand. Previously to 1853 this description of fur was principally shipped to Europe, but since that time the American furriers have been using it extensively in the manufacture of ladies' victorines, capes, &c., and the competition between the shippers and manufacturers has raised the price. The mink of Wisconsin is not so fine as those obtained from Buffalo. The best mink skins are those caught in the north and north-east of Michigan, the north-west furnishing the next quality. The range of price is very great; colour, size, and quality of fur, as well as the condition in which the skin is cured, all contributing to determine the value.

In 1851 we imported 191,729 mink skins; in 1860, 111,926. The imports by the Hudson's Bay Company range from 60,000 to 70,000 per annum and have occasionally reached 76,200. This skin at present is so highly appreciated that it commands ten times the price it did a few years ago. Very fine dark-coloured specimens sometimes sell in America for £1 each, and even higher, when manufactured into caps, tippets, &c. As in most fur-bearing animals the skin of the northern mink is the most valuable, and the fur only good when taken late in autumn, in winter, or early in spring.

In the Museum Collection of furs there are specimens of pine marten, sable, mink, ermine, cross ermine, skunk, perivitski, musquash, and others.

FISHER (*Mustela Americana*, Turton; *Mustela Canadensis*, Lin.). —Large numbers of these skins are every year exported by the fur traders from North America under the name of fisher or pekan, they are also locally called wood-shocks. The animal has a range extending completely across the American continent. The tail forms a large portion of the value of the skin in Russia, being worn as a circlet round the head. Its fur is harsher than that of the pine marten, and much less valuable, the musky odour of the body is rather stronger. The fur towards the roots is fine and downy, of a clove brown colour.

THE PINE MARTEN (*Mustela leucopus*, Lin.) inhabits the woody districts in the northern parts of America, from the Atlantic to the Pacific, in great numbers. The fur, of a dull greyish brown, is used for trimmings. Hence it has always been an important article of commerce. About 100,000 to 200,000 skins have long been collected annually in the American fur countries. The lustre of the surface of the fur is considerable, and the fur is in the highest order in the winter time.

PINE MARTEN (*Mustela leucopus*).

This is the fur generally understood in England as "sable," Russian sable being always separately indicated. The quantity sold here is very small.

Its geographical range is between 40° and 70° N. across the continent. The imports by the Hudson's Bay Company are about 5,000 to 6,000 annually, the largest number was 7,185 in 1866.

THE BAUM OR WOOD MARTEN (*Mustela abietum*) is found in the pine forests of northern Europe, where it lives like a squirrel upon the trees, which it easily climbs, its long claws giving it a firm hold upon the bark. Its fur is in the natural state coarser than that of American sable, but when dyed the skins resemble the best sable, and form a large article of commerce.

THE BEECH OR STONE MARTEN (*Martes albogularis, Mustela saxorum*) is also a European species, common in mountainous countries. The French excel in dyeing this fur, whence it is frequently called French sable.

THE KOLINSKI OR TARTAR SABLE (*Mustela siberica*) is procured from the Russian territories. Its natural colour is a bright yellow, in which state it is much used; it is also dyed to imitate the dearer sables.

THE RUSSIAN SABLE (*Martes zibellina*) is one of the most costly furs, both from its fineness, and the difficulties of procuring it amid the wilds of Siberia; it is rarely used except by the wealthy. Russia receives about 25,000 skins annually.

THE OTTER (*Lutra vulgaris*) was at one time common in England, and is still frequently caught in Europe. In China it has been trained to catch fish for its owner. The handsome fur of the otter meets with a ready sale in commercial circles. In 1871, 1,550 land otter skins were imported at the Chinese ports of Tientsin and Chiukiang.

THE AMERICAN OTTER (*Lutra canadensis*) differs from the European in its greater size, and the colour of its fur, which is fine and thick; a good skin is worth 30*s*. or 40*s*. It is used for the finer sorts of hats or for costly caps. The hunting season is from September to May. The Russians, Chinese, and Greeks use a great many of these skins for robes, trimmings, and national dresses. The following have been the annual imports by the Hudson's Bay Company during the last twenty years:

1856	13,740	1866	18,534
1857	11,577	1867	14,236
1858	12,500	1868	12,026
1859	13,165	1869	12,500
1860	11,278	1870	10,900
1861	13,199	1871	13,100
1862	14,158	1872	13,700
1863	13,331	1873	11,200
1864	15,443	1874	8,300
1865	13,600	1875	13,000

The fur very much resembles that of the beaver, but it is shorter, and not so well adapted for felting. In the collection of furs of the Museum are skins of the otter and "pulled otter," with the long hairs removed.

THE SEA OTTER (*Enhydra marina*).—The fur of this animal is exceedingly fine and heavy. It was formerly found in thousands on the Pacific coast from California to the Russian settlements, but from the absence of proper restrictions, is progressing towards extinction on those parts of the coast under British rule. The imports by the Hudson's Bay Company range from 100 to 300 per annum. In the last quarter of a century about 100,000 sea otter skins have been sent to Russia from the Pacific coast. The flesh of the young sea otter is said to be very delicate food, not unlike lamb.

The sea otter has been found only in the North Pacific. Their skins have ever been held in high estimation by the Chinese and Russians as an ornamental fur; but their great scarcity and consequent cost limits the wear to the wealthy and higher classes only. The commercial value of the sea otter's skin, like other commodities, has varied with the changes in the relation of supply and demand. The narrative of Cook's Voyage shows the value of a prime skin to have been at the time of that voyage £24. In 1802, when the largest collection was made, the average price of all skins at Canton was only about £4. The skins were formerly worth in Europe from £40 to £100, but they have much declined in value. At present the best quality only bring from £16 to £20.

When properly skinned the pelt is of an oval form. The tails are always cut off and sold separately. The hair in a first-class sea otter skin, according to Mr. Dall, should be nearly even in length all over it, and of uniform colour. The length of a full-sized skin which has been stretched before drying, is about six feet, and its breadth nearly four feet.

The fur varies in beauty according to the age of the animal.

When the otter is full-grown it becomes of a jet black. The male otter is beyond all comparison more beautiful than the female, and is distinguished by the superior jetty colour, as well as velvety appearance of his skin. The skins in the highest estimation are those which have the belly and throat plentifully interspersed with a kind of brilliant silver hairs, while the body is covered with a thick black fur of extreme fineness and silky gloss.

The sea otter possesses historic interest. Its quest led the Russians from Okhotsk to Kamschatka, and thence over the Aleutian chain to the opposite coast of America. As long ago as in 1788, 3,000 skins were obtained by Delouff in Cook's Inlet; and in 1794, Barnoff's expedition to Behring Bay took away about 2,000. The animal was formerly abundant on the Kamschatkan coast, and on the American coast as far south as Lower California; but owing to its habits it is difficult to prevent indiscriminate slaughter, and it is now scarcely known in many localities where it was once abundant. We have no accurate information of the total quantity taken annually.

SEALS.—In a commercial point of view, the seals are the most important of carnivorous animals, and the pursuit of them is eagerly carried on in the Arctic and Antarctic regions, as well as in other quarters, but is most extensively prosecuted around the shores of Newfoundland and the Labrador coast, for the skins and oil obtained from the animals.

The various kinds of seals differ as greatly in size and physiognomy as members of the human family. There are canine and feline-looking seals; seals with round heads cropped like a prizefighter's, and seals with patriarchal beards and flowing locks; meek, pensive-looking seals, and seals fierce and long tusked; little seals three feet in length, and monsters upwards of eight feet long, weighing 1000 lbs. The harp seal is most esteemed, and commands a high price. The Greenland seal is double the size of the common seal, and the sea-lions range from 18 to 20 feet in length.

The seal fishery is now the most profitable branch of trade in the colony of Newfoundland. In 1852 it employed 367 vessels, of 35,760 tons aggregate tonnage, and 13,000 men. The vessels are from 75 to 200 tons, but those of 130 tons, which carry crews of 40 or 50 men, are preferred. The voyage is begun early in March, rarely exceeds two months, and is often completed in three weeks. Two and three voyages are sometimes made in one season.

Assuming the fleet of vessels fitted out in the colony at 360, they average over 100 tons, and employ an aggregate of 15,000 men. Pricing the vessels at £1000 apiece, we have £360,000 (irrespective of provisions) of floating capital on the waters. In the brief space of eight years the whole of this capital disappears, for the fleet wears out within that period, and has to be replaced. Hence £40,000 a year (besides the cost of annual repairs) is expended for a portion of the decked vessels with which the fisheries are carried on from Newfoundland alone. It may be remarked that of late years the number of vessels is fully 100 below what it used to be. In 1860, 200 vessels took 258,015 seals; in 1870, 116 vessels took 189,733.

It is the opinion of those who best understand the business, that the fleet despatched annually is already too extensive for the return that can fairly be expected from the ice-fields in such proximity to the coast of Newfoundland, as will render an outfit from its ports convenient or profitable. The exports of seal-oil, during the five years ending with 1850, averaged 4,921 tuns for each year; for the five years ending 1855, 6,353 tuns; five years ending 1865, 3,538; five years ending 1870, 4,900 tuns; but the skins also yield very large profits.

It is stated that the prosecution of this "fishery" is hazardous and very exciting. When the voyage is successful, the gain to the merchant, captain, and crew is very considerable, but the losses are also heavy. It places large sums of money in the hands of ordinary seafaring men, who squander it characteristically.

Governor Darling says it is a pursuit which partakes more of gambling than of steady industry, and he is of opinion that it does not improve the physical condition or promote the happiness of the people at large.

The species of seal which chiefly resort to the Newfoundland coast are the two largest kinds—the hooded seal (*Stemmatopus* [*Cystophora*] *cristatus*), 7 or 8 feet long, and the harp seal (*Phoca Greenlandica* and *oceanica*). They whelp their young, in January and February, on the ice-fields of Labrador. The whelping ice, as it is called, is floated southwards by the ocean-currents, and is always to be found on the coast of Newfoundland after the middle of March. The young seals, not taking to the water till three months old, are easily caught; their skins, with fat attached, are stripped off, and the worthless carcases left on the ice. They are sorted into five qualities—"young harp," "young hood," "old harp," and "bedlamer" (a year-old hood), and "old hood," the most productive being "young harp."

At St. John's, the headquarters of the trade, the skins and blubbers are separated, and the latter is generally put into wooden cribs, beneath which are pans to catch the oil. No artificial heat is used in this process. The oil which runs for the first two or three months is termed *pale seal oil*, and forms 50 to 70 per cent. of the whole quantity. As putrefaction takes place, the oil becomes darker and more offensive. The putrescent refuse and the clippings of the pelts yield further quantities of oil by boiling (*boiled seal oil*). This old process, however, is now superseded by a steam apparatus. By this invention a uniform and much better quality of oil is produced, free from the horrible odour of that prepared by the old method, and the time required is only twelve hours, instead of six months.

The aggregate annual value of the seal-oil and seal-skins shipped from Newfoundland ranges from £250,000 to £300,000.

The average take of successful vessels is about 2,000 seals, though as many as 8,000 have been secured in one trip; out of upwards of 400 vessels that yearly engage in sealing, not more than sixty make remunerative voyages, and many suffer heavy losses. Hence the business is altogether a lottery. The seal fishery of Newfoundland for 1876 was, however, one of the best for many years, and the following is a graphic account of the results:

"For seven weeks the bitter east wind blew, driving the huge ice-fields upon our shores, filling all the bays and harbours with ice, and creating a solid ice-pack, which no vessel could penetrate, all around the coasts. The sealing fleet were arrested within sight of the harbour and held powerless in the grasp of the frost king. Thus matters stood, and we anticipated a disastrous fishery. Our hunters could not get to the seals, lying in their icy cradles, rocked by the heaving billows of the Atlantic, but all the while the east wind was gently floating the 'white coats' within reach of the hunters, till at length they were separated from detachments of them only by a few miles. A change of wind came at last, a relaxation of the ice took place, and the hunters sprang upon their prey. Still they did not come up with the great body of the seals, only with scattered herds, so that no steamer got a 'bumper trip,' though many did well on their first voyage. The largest number brought in was by the steamer Bear—20,000 seals, worth three times as many dollars. Others got 16,000, 13,000, and 10,000 each. The most lucky hit of all was made by the Merlin, an old worn-out steamer. She was pinned in near the shore by the ice pack, but the seals came around her, and she got as many as she could load—16,000 prime seals, value 48,000 dols. The seals this year were not taken till in their full-grown stage, and consequently were on an average 20 per cent. better than in other years. So soon as the steamers unload their cargoes of fat they start again for the ice-fields on a second hunting excursion. On this trip they rarely capture many young seals, as they have taken to the water about the 1st of April; but they pursue the old seals, sometimes shooting them on the 'pans' of ice, sometimes falling in with a herd of them jammed in ice and unable to escape, and 'batting' them in multitudes. This year will be memorable in the annals of seal killing by the wonderful success of second trips and the vast number of old seals brought in. Let us follow the fortunes of the steamer Neptune as an illustration of the romantic side of seal killing. She did but moderately well on her first trip, having brought in only 8,000 seals, value 24,000 dols. On her second trip she got caught in an ice pack in Green Bay, and vainly tried to escape. When there the men were sent out over the ice fields, and some of them wandered so far as ten miles from the steamer. At this spot a huge herd of old seals was discovered caught in the pack and unable to escape. The men returned to the ship with the welcome news, and the whole crew, 30 in number, formed themselves into line like soldiers charging, and rushed on their prey. The work of destruction by striking them on the nose with a long club called a 'gaff' was eagerly pursued, and at the end of four hours 18,000 great seals lay dead within a small area. But how were they to be got aboard, the vessel being ten miles distant? To drag them over the ice was impossible. Well, by one of those rare pieces of good fortune which sometimes befall the hunter, the grasp of the ice relaxed—the east wind ceased to press from the outside, and next morning the steamer was able to get alongside the slaughtered seals and all were easily put on board. The striking thing, too, was, that had they not been killed on that particular afternoon the whole would have escaped next morning through the openings in the ice. The average worth of an old harp seal is 6 dols.; so that the value of this cargo was 108,000 dols. This is the most valuable cargo of seals ever taken in Newfoundland. This single steamer thus earned 132,000 dols. in a little over two months. One third goes among the men; the captain will get 4,000 dols. for his share; the remainder belongs to the owners. Several other steamers have arrived with good trips of old seals. The Wolf has 8,400, Ranger

7,000, Walrus 3,800, Greenland 4,300. The Vanguard and Commodore are also said to have fair trips, but are not reported. When one remembers that every old seal is worth 6 dols., the value of these united cargoes is very considerable. The weight in fat of the Neptune's cargo is 850 tuns. All the captains unite in declaring that they never saw the seals more numerous, so that to all appearance our seal fishery presents as yet no signs of exhaustion. But I should like to hear what Professor Baird

GROUP OF SEALS.

would say to this terrible slaughter of old seals, coming after the destruction of the young ones. It seems like killing the goose that lays the golden eggs. It is vain to enact restrictions when men are out in those ice solitudes and herds of seals around. Not till unmistakable signs of an exhaustion appear shall we get the killing of old seals prohibited.

"The persistent east wind drove the great body of the young seals up the bays and in upon islands and headlands on our northern coast, thus bringing them within reach of the settlers ashore. The whole population of these places rushed out and slaughtered and dragged the seals ashore. 'Young men and maidens, old men and children,' were eagerly engaged in the work. News arrived some time ago that in two localities, Twillingate and Fogo, 100,000 seals had been taken in this way—value, 300,000 dols. It is supposed that at least 50,000 seals additional must have been taken in other neighbouring localities from which no news has yet arrived. Where is the gold mine in the world that can compare with our seal fishery? How sad if by reckless destruction of these valuable creatures we should exhaust this important industry! That there is fear of such a result is evident when we look at so many other fisheries once as flourishing as ours and now non-productive."

Seal-oil is also obtained in the Caspian and White Seas, and other parts of Russia. In the Caspian Sea about 140,000 to 160,000 pouds of 36 lbs. are obtained annually; in the White Sea half that quantity. The Capuchin or hooded seal (*Cystophora cristata*) is found to yield there 360 lbs. of blubber, the *Phoca Grœnlandica* 160 to 240 lbs., the *Phoca annellata* 120 lbs., and the young white seal about 60 lbs.

Repeated and careful experiments in rendering out seal fat or blubber in Newfoundland, show the relative produce of pure oil obtained from the different varieties of seal to be as follows, per barrel, when in prime condition—

	Blubber. lbs. of	Oil. galls.	Residue. lbs.
Old harp (*P. Grœnlandica*) . . .	288	22½	73
Young harp	225	22	52
Young hood (*Cystophora cristata*) . .	230	21	80
Bedlamer, a year-old hood . . .	246	21½	103

The average quantity of oil from 1,000 seals may be roughly estimated at 10 tuns. Seal-oil has ranged in price from £30 to £32 10*s.* per tun. Seal skins according to size, from 8*s.* to 24*s.*

At the Cape of Good Hope the technical names for seals are wigs, pups, and black pups, the middling and small skins are considered the best.

Seal skins in crust are sold according to weight, those of 25 lbs. to 45 lbs. will fetch 78*s.* to 100*s.* per dozen, those weighing 50 lbs. to 80 lbs., 120*s.* to 200*s.* When split for binders they sell at 78*s.* to 90*s.* per dozen, or at from 5*d.* to 10*d.* per lb. The common terms for the different kinds of skins are, blue backs, white coats, and hair seals.

While the products of these animals have become regular articles of commerce, and contribute to the requirements of civilised life, it should also be remembered that they are even more essential to the well-being of the tribes of men inhabiting the Arctic regions. "Seals (observes Crantz) are more needful to them than sheep are to us. The seal's flesh supplies them with palatable and substantial food; the fat is sauce to their other aliment, and furnishes them with oil for light and fire, while at the same time it contributes to their wealth in every form, seeing that they barter it for all kinds of necessaries. They sew better with the fibres of seal's sinews than with thread or silk; of the fine internal membranes they make their body raiment and their windows; of the skins they make their buoys, so much used in fishing, and many domestic utensils, and of the coarser kinds their tents and their boats of all sizes, in which they voyage and seek provisions."

It has been recently stated, "Whales and walruses the Esquimaux capture when they come in the way, but the seal is their daily bread; his flesh and blubber support them and feed their lamps; his skin clothes them and forms their kayaks, with which they brave the stormiest seas; the seal's bone, where iron is not to be had, barbs their harpoons, and seal's bladder forms the float with which, when the prey is speared, it is so hampered that it is unable to escape."

There is no food more delicious to the taste of the Esquimaux than the flesh of the seal, and especially that of the common seal

(*Phoca vitulina*). But it is not only the human inhabitants who find it has such excellent qualities, all the larger carnivora prey on seals. Seal's meat is so unlike the flesh to which Europeans are accustomed, that it is not surprising that we should have some difficulty at first in making up our minds to taste it; but when once that difficulty is overcome, everyone praises its flavour, tenderness, digestibility, juiciness, and its decidedly warming after effects. Its colour is almost black, from the large amount of venous blood it contains, except in very young seals, and is,

ESQUIMAUX DOGS CAPTURING A SEAL.

therefore, very singular-looking, and not inviting, while its flavour is unlike anything else, and cannot be described except by saying "delicious!" To suit European palates, there are certain precautions to be taken before it iscooked. It has to be cut in thin slices, carefully removing any fat or blubber, and is then soaked in salt water for from 12 to 24 hours to remove the blood, which gives it a slightly fishy flavour. The blubber has such a strong taste that it requires an arctic winter's appetite to find out how good it is. The daintiest morsel is the liver, which requires no soaking, but may be eaten as soon as the animal is killed. The heart is good eating, while the sweetbread and kidneys are not to be despised. The usual mode of cooking seal's meat is to stew it with a few pieces of fat bacon, when an excellent rich gravy is formed, or it may be fried with a few pieces of pork.

The Esquimaux use every part of the seal, and, it is said, make an excellent soup by putting its blood and any odd scraps of meat inside the stomach, heating the contents, and then devouring tripe, blood, and all with the greatest relish. "For my own part" (observes Dr. Horner) "I would sooner eat seal's meat than mutton or beef, and I am not singular in my liking for it, as several of the officers on board the Pandora shared the same opinion as myself. I can confidently recommend it as a dish to be tried on a cold winter's day by those who are tired of the everlasting beef and mutton, and are desirous of a change of diet."

In the Museum Collection, besides a white coat seal skin in the natural state, and one dyed, and the stages of preparation of seal skin for leather, there is a model Caiak or Esquimaux sealing canoe with all its harpoons, weapons, inflated float, and appliances; different kinds of seal oil, pale, straw, brown, &c. Porpoise oil and sea elephant oil are among the animal oils.

The following gives the annual average exports of seal oil from Newfoundland at quinquennial periods :—

	Tuns.		Tuns.
1850	4,291	1865	3,538
1855	6,353	1870	4,901
1860	5,816	1874 (4 years)	4,332

SEAL SKINS IMPORTED INTO THE UNITED KINGDOM.

1840	560,596	1849	470,834
1841	313,362	1850	779,924
1842	153,828	1851	769,756
1843	772,697	1852	811,530
1844	563,947	1853	850,550
1845	632,304	1854	661,552
1846	428,633	1855	601,002
1847	753,141	1856	681,234
1848	706,267	1857	803,438

1858	719,926	1867	743,511
1859	565,813	1868	780,477
1860	561,666	1869	736,336
1861	494,079	1870	731,913
1862	480,526	1871	833,709
1863	555,334	1872	657,696
1864	342,833	1873	876,077
1865	529,284	1874	755,005
1866	513,671	1875	629,723

THE FUR SEALS OF COMMERCE.—The fur seals and sea lions are closely allied, forming the family *Otariadæ*. They are well distinguished from the hair seals (*Phocidæ*) by their external ears and long flippers destitute of hair, and with only three nails. The hair seals have no external ears, and their flippers are broad, short, and covered with hair, having five nails on the hind ones. For many years great ignorance prevailed as to the animals which furnished the true fur seal-skin of commerce, and even yet, professed naturalists admit there is a lack of knowledge with regard to them, their natural history having been studied, for the most part, only in a limited and fragmentary way, scarcely a single species having been fully investigated. Millions have been slaughtered for their skins, but those engaged in the murderous traffic care nothing for the scientific characters of their victims. Naturalists of exploring expeditions have made isolated observations and brought home a few skins and skeletons; but for connected and extended studies of these commercially valuable and scientifically interesting animals, we are mainly indebted to the report of Mr. Elliott, who was sent out by the American government to investigate the sealing and other resources of their lately purchased islands in the North Pacific, and to the observations of Captain Musgrave, who beguiled the tedium of a twenty months' enforced sojourn on the Auckland Isles by watching the species inhabiting that region.

Mr. A. W. Scott also published in 1873, at Sydney, an elementary treatise on "Mammalia recent and extinct," confined however to the Pinnata, Seals, Dugongs and Whales. In the preface he tells us "influenced, by the great commercial value of several species of the pinnata, I have felt anxiously desirous to direct, without further delay, the attention, and thus possibly secure the sympathy, of readers, other than students, to the necessity ot prompt legislative interference, in order to protect the oil and fur producing animals of our hemisphere against the wanton and unseasonable acts committed by unrestrained traders; and thus not only to prevent the inevitable extermination of this valuable group, but to utilize their eminently beneficial qualities into a methodical and profitable industry.

"Keeping steadily in view these two objects, I have endeavoured by devoting as much space as my limits would permit to the consideration of the animals whose products are of such commercial value to man, and whose extinction would so seriously affect his interests, to point out the pressing necessity that exists for devising means of protection for the fur seals and the sperm and right whales of the Southern Ocean.

"To evidence what great results may be effected by considerate forethought, I may state that under the fostering care of the United States Government, the northern fur seals of commerce, which but a few years ago were nearly extinct, have already, by their rapid increase and mild disposition, developed themselves into a permanent source of national wealth.

"The islands of the Southern Seas, now lying barren and waste, are not only numerous, but admirably suited for the production and management of these valuable animals, and need only the simple regulations enforced by the American Legislature to resuscitate the present state of decay of a once remunerative trade, and to bring into full vigour another important export to the many we already possess."

There are four classes of seals on the Prybilov Islands (Behring's Straits): the fur seal (*Otaria* [*Callorrhinus*] *ursinus*), the sea lion (*Otaria* [*Eumitopius*] *stellerii*), the hair seal (*Phoca vitulina*), and the walrus (*Trichecus rosmarus*). The flesh of the hair seal is stated to be more juicy and sweet for food than that of the fur seal.

A most instructive report was issued by Mr. Elliott, Assistant Agent in the Treasury Department, who was deputed a couple of years ago to examine the Prybilov Islands of Alaska, the great rendezvous of the sea bears, as they are termed, which gives hints for the practical management of these lucrative fisheries for the future. Mr. Elliott informs us that the finest sorts of the true fur seals repair to these Islands annually to breed, in fabulous numbers—that but few members of the animal kingdom exhibit a higher order of instinct and intelligence —and, that the male sea bear is, of all brute polygamists, the most notorious.

The adult male is from 5½ to 7½ feet in length, and weight about 400 lbs. Millions of them may be seen at once on some of their "rookeries," or breeding grounds, along the coasts of the rocky islands they frequent. In the month of June the bull seals arrive in thousands, and the females come up out of the sea in still greater numbers about three or four weeks later. Some of the bulls display wonderful strength and courage. They swim very swiftly, and are as great a terror to the smaller species of seals, such as the *P. vitulina* and the like, as the great sea lion is to them. Their skin, which is very thick, is covered externally with hair like that of the common seal, but a good deal longer, and standing erect about the shoulders of the male. Beneath this external hair, the animal's body is clothed in the soft wool, which, in a manufactured state is familiar to us all; and is such a large source of revenue to the London and Continental furriers in the shape of ladies' seal-skin cloaks, jackets, trimmings, and comfortable articles of wear for the winter.

Sir George Simpson, who had many opportunities of studying the habits of the North American animals in their native haunts, speaking of the fur seal in former times, says:—"Twenty or thirty years ago there was a most wasteful destruction of the fur seal, when young and old, male and female, were indiscriminately knocked on the head. This improvidence, as every one might have expected, proved detrimental in two ways. The race was almost extirpated and the market was glutted to such a degree, at the rate of some 200,000 skins a year, that the prices did not even pay the expenses of carriage. The Russians, however, have now adopted nearly the same plan which the Hudson's Bay Company pursues in recruiting any of its exhausted districts, killing only a limited number of such males as have attained their full growth—a plan peculiarly applicable to the fur seal, inasmuch as its habits render the system of husbanding the stock as easy and certain as that of destroying it."

In Mr. Dall's work, "Alaska and its Resources," the Arctic fur seal trade is fully detailed, and the numbers caught seem to have been very large, as will be seen by the following figures:—

Caught	1786 to 1797	557,024
Exported	1797 to 1821	1,767,340
Sold in the Colonies	ditto	377,642
Caught	1821 to 1842	758,502
Caught	1842 to 1862	372,894
	Total in 76 years	3,833,402

Or an average of more than 50,000 a-year.

The contract for taking fur seals on the islands of St. Paul and St. George, in Alaska, was a few years ago awarded by the American Congress to the Alaska Commercial Company, of San Francisco. This company is to have the privilege for twenty years, on payment of 55,000 dollars annual rental, 62½ cents on each skin taken, and 55 cents for each gallon of seal oil obtained.

The report of Lieutenant Washburn Maynard, of the United

States Navy, who was detailed to accompany Mr. Elliott to Alaska has been laid before Congress. Lieutenant Maynard states that the Alaska Company are observing their contract, 100,000 seals are annually killed by them, and 12,000 by the natives for food, under the direction of the Government agent. The number is experimental, and Lieutenant Maynard is of opinion that the effect upon the increase or decrease of seal life upon the islands cannot be discovered for six or seven years from the making of the contract. He gives a chapter on the sea otter, the skin of which animal is ten times as valuable as the seal. He reports, as do others, that this animal is rapidly disappearing by the indiscriminate slaughter of the female and young otter, the constant harassing by hunters, and the use of fire-arms in their capture, and he indicates that prompt measures must be taken, or it will be too late to preserve the otter from extinction.

Mr. J. Willis Clarke gave a Lecture on Sea Lions before the Zoological Society, which was published in the "Contemporary Review" for Dec. 1875.

According to this naturalist, there are nine well authenticated species of sea lions, thus distributed: in the North Pacific three, *Otaria* [*Callorrhinus*] *ursinus*, *O. Gillespii*, and *O. stellerii*; in the South Pacific, around Cape Horn, and in the South Atlantic as far north as Rio de la Plata, two species, *O. jubata*, and *O. Falklandica*; about the Cape of Good Hope and the adjacent Islands, one, *O. pusilla* or *antarctica*; around Australia and New Zealand, two, *O. Hookeri* and *O. lobata*; and at Kerguelen's Land, one, *O. gazella*.

SEA LIONS (*Eumitopius* [*Otaria*] *stelleri*, Mull.).—These animals are abundant on most of the rocky islands of Alaska. They appear in May, and remain until late in the fall. The males often weigh two or three tons. Their hide and oil are used for the same purpose as those of the walrus, though inferior in quality. The whiskers of the sea-lion are as large as a quill, and sometimes fifteen inches long. They are exported to China, the Chinese paying a high price for them to use as toothpicks. The gall is also disposed of in China, being used in the manufacture of silk.

The flesh of the fur seal and sea-lion serve the Aleuts for food, and their blubber for fuel. The flesh of the fur seal forms but a small portion of the body; the greater part is blubber, and this is more noticeable because of the thousands of carcases of seals which are scattered over the Aleutian Islands. If these were composed of a large part of muscular fibre, as is the case with the walrus, the decaying bodies would breed a pestilence. The flesh of a young fur seal, placed in running water overnight, and then broiled, is far from disagreeable—in fact, it tastes exactly like mutton-chop. The young sea-lion is said to be even better eating, and both present a marked contrast to the fetidity of the flesh of the hair seal (*Phoca*) of Norton Sound. The Aleuts make of the skin of the flippers boot soles, which are very durable.

The fat cut from the nearest carcase serves them for fuel. The blubber of the fur seal yields oil of the first quality, and is worth about two dollars a gallon; yet for many years hundreds of barrels have fertilised the hillsides, for want of some one to preserve it. Each seal will furnish half a gallon, which would give, for 100,000 seals, about 1,000 barrels of oil, worth at least £12,000; this has always been wasted. In fact, the oil is worth as much as the skins at the islands.*

Sea lions are unknown in the Atlantic except in the extreme south, though the Atlantic abounds in true seals, from which the sea lions differ in several particulars. The more obvious difference is the possession of external ears, which seals lack. They have besides a long, mobile, flexible neck, whereas in seals the neck is short and scarcely perceptible. Then their limbs are still available for locomotion on land, while those of

* Dall's "Alaska and its Resources."

seals have lost all power of supporting the body out of water. Lastly, they possess the fatal gift of under fur, which gives them their commercial importance, and threatens to cause their untimely extermination.

From time immemorial two species (*Otaria* [*Eumitopius*] *stellerii* and *O. Gillespii*) have inhabited an island in San Francisco Harbour. Protected by the civic authorities, they have multiplied enormously, threatening the entire destruction of the salmon once so plentiful in all Californian rivers.

NORTHERN FUR SEAL, THE SEA BEAR OR SEA LION OF ALASKA.

Of the Arctic or Behring Sea species (*Callorhinus ursinus*, Gray) not less than six million skins have been obtained since 1741. The Hudson's Bay Company received between 1856 and 1866 upwards of 4,000.

They are killed by a blow on the back of the head with a heavy sharp-edged club. The Aleut then plunges his sharp knife into the heart, and with wonderful dexterity, by a few sweeps of his long weapon, separates the skin from the blubber to which it is attached. The nose and wrists are cut around, and the ears and tail left attached to the skin. When the operation is over the skin is of an oval shape, with four holes where the extremities protruded. These skins are then taken and laid in a large pile, with layers of salt between them; after becoming thoroughly salted they are done up, two together, in square bundles, and tied up with twine. They are then packed for transportation to London, where all the fur seals are dressed. (Dall's "Alaska.")

The claim of the North Pacific sea lions to public interest arises from the circumstance that eighty per cent. of the seal fur now supplied to the markets of the world comes from the islands of Behring's Sea, and the indications are that in a few years the Alaskan possessions of America will be the only source of this beautiful fur. Everywhere else the slaughter goes on without regard to system, age, or sex, and already many islands which used to furnish thousands of skins every year have been entirely depopulated. Not less than a million skins were taken from the Island of Masafuera, off the coast of Chili. In two years, four hundred thousand skins were obtained from a small island near Australia. From the South Shetland Islands, three hundred and twenty thousand were taken in 1820 and 1821, males and females being slaughtered indiscriminately, and the young left to die. It is hardly necessary to add that, in a few years, this horrid and wasteful process wrought its own destruction.

The resorts of the sea lions of the north were discovered in 1786, and a Russian fur company at once established. For thirty years from eighty to ninety thousand skins a year were brought away, the killing being done without regard to sex or system. About 1817 it was observed that the animals had diminished in numbers. For twenty years more the wasteful slaughter went on, until but a tithe of the original number appeared. Then the slaughter was regulated, the number of skins restricted, and the females left undisturbed. When the islands came into the possession of the United States, the system was substantially continued, with the result of giving them almost a monopoly of the entire seal fur trade. According to Mr. Elliott's calculation, as many as three million breeding seals annually congregate on the two islands, St. Paul and St. George, to which they resort; the yearlings and males under six years of age he sets down at two millions, making a population of upwards of five millions. Only young males are allowed to be killed, and the number is limited to one hundred thousand. Females are not molested, and no killing is permitted within several miles of the "rookeries," as the resting places of the females and their cubs are called.

When the time for killing arrives, usually in June, the killers select some "hauling ground" of the young males—for the old bulls do not allow them to associate with the females—and, armed with clubs, get between them and the sea. The animals, startled by the sight of the men and frightened by their shouts, scramble rapidly landwards, and are leisurely driven to the killing grounds. In favourable weather they can travel at the rate of half a mile an hour, the most effective implement for driving being an umbrella. At the killing ground they are allowed to rest awhile, after which the fittest are selected and killed with clubs, a single blow on the head being sufficient for each. The rest are allowed to return to the sea.

For many years the stiff coarse hair, which conceals the under fur, was plucked out by hand. It was finally discovered that the roots of the hair were more deeply seated than those of the fur, and that, by shaving the skin from the underside until the hair roots were cut away, the hair could easily be brushed away, leaving the under fur intact, thus greatly simplifying and cheapening the work of preparing the skins. Naturally, the under fur is curly and of a light brown colour; but as the ladies prefer a darker shade, it is dyed, in which process the curls untwist and the fur becomes smooth. The mode of preparation is described at p. 211.

The Government rents the islands for 50,000 dollars a year, and imposes a revenue tax of two dollars on each skin taken. According to late accounts, the seal population of the islands is steadily increasing, and it is considered that the number allowed to be killed might safely be increased also. In view of the probable early extermination of the fur-bearing seals, so called, of other regions, however—at least, so great a reduction of their numbers as to make the taking of them unprofitable—it is to be hoped that no risks will be run in the only place where they have any chance of perpetuation. Better under-kill than over-kill, even if the demands of the ladies should be scantily gratified. Properly managed, the Alaskan Islands will remain for ever the chief source, perhaps the only source, of this beautiful and valuable fur.

When full grown, a sea lion is about 15 ft. long and weighs 16 cwt. The chief home of these seals is the ocean and shores between Russian North America and the opposite shores of Russia itself—that is to say, about Behring's Straits and Behring's Sea. They are found also in the Kurile Islands and east coast of Kamschatka.

The sea lions of the Southern Ocean are the *Otaria jubata* and *O. Falklandica*. These have been pursued with an indiscriminate slaughter in the high Antarctic latitudes by the seal hunters who annually go south to the Crosets, Kerguelen Island, and other desolate places for skins and oil; instead of only destroying a proportion of the adult males, they put to death the females also, and the helpless cubs perish from cold and hunger alongside the dead bodies of their mothers.

The Antarctic fur seal (*Arctophoca Falklandica*, Peters, *Otaria Falklandica*) was at one time common in the Falkland group and the adjacent seas. The skins (which were worth fifteen Spanish dollars, according to Sir John Richardson) are from four to five

THE SEA LION (*Otaria Falklanaica*) OF THE ANTARCTIC SEAS, ONE OF THE FUR SEALS OF COMMERCE.

feet long, covered with reddish down, over which stiff grey hair projected. They were especially hunted on the Falkland Islands, Terra del Fuego, New Georgia, South Shetland, and the coast of Chili. Three-and-a-half million of skins were taken from Masafuera to Canton between 1793 and 1807 (Dallas).

THE SEA ELEPHANT (*Macrorrhinus angustirostris, Phoca proboscidea*, Peron) of California has only been well described since 1866. The males have a sort of small trunk, but no tusks. Its flesh is not only black, oily, and indigestible, but it is also almost impossible to separate it from the lard. The tongues alone supply really good aliment, and they are salted with care and esteemed in the market. The heart is sometimes eaten, but it is hard and indigestible; and with regard to the liver, which is esteemed in some seals, according Dr. Hamilton, it would appear, after repeated trials, to be hurtful.

WALRUS (*Trichecus rosmarus*), SHOWING THE UPPER INCISORS IN THE FORM OF TUSKS.

THE WALRUS OR SEA-HORSE (*Trichecus rosmarus, Rosmarus obesus*) it has been well remarked, forms a connecting link between the mammalia of the land and those of the water, corresponding in some of its characters both with the bullock and the whale. It is often seen of the size of a great ox, and sometimes exceeds the dimensions of the gigantic elephant. The chase of the walrus in the Arctic regions is of great antiquity. They used to congregate by thousands on the Magdalen Islands in the River

St. Lawrence, but have long been driven far to the north, by their pursuers.

The economic products for which this animal is sought are its flesh and its skin, its oil and teeth. Among the inhabitants of the Arctic regions its flesh is much valued and esteemed, and is greedily eaten along with the lard, and even the skin.

In ancient times most of the ropes in the vessels of northern countries, appear to have been made of walrus skin, and when cut into shreds and plaited into cordage, it formed lines which were used for the capture of the whale; these also answer admirably for tiller-ropes; cables, too, were wont to be manufactured from them, and the Finlanders used to pay tribute to the king in

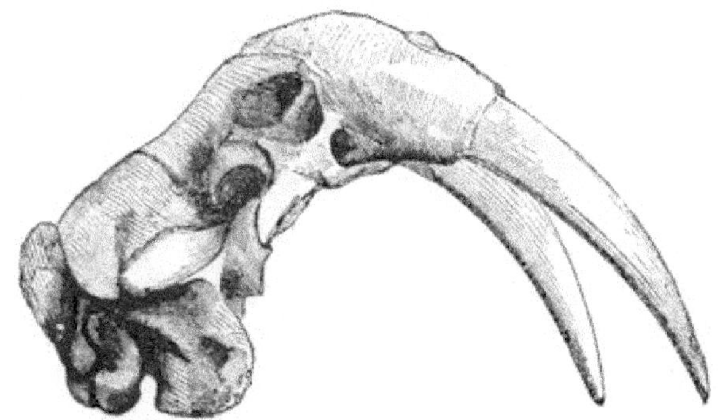

WALRUS SKULL, SHOWING THE POWERFUL CANINE TEETH.

this form. When tanned the skin is converted into a soft porous leather, above an inch in thickness, but it is not so useful nor so durable as in its green or raw state. It might do for harness and carriage leather. The hide has been successfully used for belting, and covering skin boats.

A recent American paper states that "probably not less than fifty thousand walrus, with their young, were killed and destroyed last year (1875) by our arctic whalemen. The arctic walrus never forsake their young, but will take them in their flippers and hold them to their breasts, even when their destroyers are putting their sharp lances through and through them and the blood is streaming from every side, uttering the most heartrending and piteous cries until they die. But the worst feature of the business is, that the natives of the entire Arctic shore are now almost entirely dependent upon the walrus for their food, clothing, boots, and dwellings.

WALRUS ON THE ICE.

Twenty years ago whales were plentiful and easily caught; but they have been driven north, so that now the natives seldom get a whale."

These animals are abundant near Port Mollis in Bristol Bay, and on the more northern coast and islands of the Alaska territory. The oil is valuable, but they seldom yield more than 20 or 30 gallons. The teeth, which weigh about 4 lbs. the pair (although many books give them at half-a-ton !), used to be in great demand by dentists. Among the Chinese the dentine or ivory is employed for those curious uses to which they turn ivory so skilfully. This oil is a well known article of commerce. There is no doubt but that the annual supply might be largely augmented. The quantity of walrus tusks obtained in Alaska averages 100,000 lbs. in weight. These canines are sometimes two feet long. The Hudson's Bay Company occasionally import about 100 lbs. or 200 lbs. weight in the year.

There are some stuffed heads of the walrus on the east wall of the Museum, one male, and two female; samples of walrus hide in Case **140** and the skin prepared as chamois leather.

CHAPTER X.

MAMMALS OF SECONDARY IMPORTANCE.

We diverge in this chapter from the higher class of mammals to notice some few which are of minor interest in an economic point of view. These are monkeys, bats, insectivorous and toothless animals, such as the hedgehog, the tatoo, the porcupine, ant-eater, and the pangolins or scaly ant-eaters. The rodent or gnawing animals are, however, more extensive and important, having regard to their commercial products, many of them yielding furs, while others, such as hares and rabbits, contribute largely to the food of man.

LEAVING the Carnivora, we have now to bestow a passing glance at a few orders of mammals, which have but a limited commercial or economic value, but still can scarcely be passed over altogether without incidental notice. These are the Quadrumana, the Cheiroptera, the Insectivora, and the Edentata ; the Rodentia will require a more detailed description.

MONKEYS.—In an economic point of view the Quadrumana are not of much importance. Live animals, it is true, are in request for Zoological Gardens, and the skeletons and skulls of some of the rarer species, as the *Troglodytes Gorilla*, realize a high price. Attempts have been made to introduce monkey skins into commerce as furs, and specimens of these black and grey monkey skins are shown in the Museum Collection. The long shaggy black skin of one species, the white-thighed Colobus, with a silky fur, obtained from the Gaboon, West Africa, when first introduced met with a very dull sale. The price at which it could be bought, however, encouraged speculators, and the skin was largely manufactured into muffs, which from the low rate they were offered at, and the really good appearance they presented, did not fail to receive a favourable reception from the ladies.

A fashion arose, and from the original price of 1*s.* per skin, they advanced in the year 1860 to 12*s.* per skin. Imitation by another long-haired fur, and less care in manufacture, occasioned the skins again to recede, and will afford time to the monkeys, (who must have been sorely hunted,) to recruit their numbers. There are two other species of monkey which might be intro-

a. Head of Aye-Aye of Madagascar (*Cheiromys Madagascariensis*).
b. Head of White-necked Marmoset (*Jacchus bicolor*).
c. Head of East Indian Red Monkey (*Macacus rufescens*).

duced with advantage; they are the Diana monkey and the *Colobus guereza*, found in Abyssinia, which is black but covered with a kind of mantle of long white hair, this, parting on the back, covers all the body. Monkeys are more extensively eaten as food than is generally supposed, in the Eastern Archipelago, Africa, and South America. In some places monkeys have been trained to ascend the palms and throw down the cocoa-nuts.

In the Museum collection there is a stuffed specimen of a small ourang-outang and a necklet of monkeys' teeth.

We give an illustration showing the facial varieties of some few different monkeys (p. 289).

BATS.—Specimens of the large edible flying fox or bat (*Pteropus edulis*), which is regarded as a delectable food by the natives of Siam, were shown at the Dublin Exhibition of 1865. They are caught in bags at the end of a pole. The flesh is said to be white, delicate, and remarkably tender, and like the flesh of most bats is generally eaten throughout the islands of the Eastern Archipelago, as is also that of the fox monkey or flying lemur (*Galeopithecus volans*).

In caverns in different countries there are large deposits of bat guano, which are utilised for fertilising land. In numerous caves from Virginia to Texas are found deposits of this excrement, sometimes reaching to many thousands of tons in extent. The Governor of the Bahamas reports that in caves in those islands there is probably 400,000 tons of this manure.

There is a skin of the vampyre bat (*Phyllostoma spectrum*), the dangerous bloodsucker of the South American tropics, among the fur series in the Museum collection; also one of the common mole (*Talpa Europæa*), the skins of which have some small value.

HEDGEHOG (*Erinaceus Europæus*).—The flesh of this animal when it has been well fed is said to be sweet and well flavoured, and is eaten in many places on the Continent. But in Britain few besides the gipsies partake of it. The prickly skin appears to have been used by the Romans for hackling hemp, but teasels and metallic "cards" have long supplanted it. It is used sometimes as a muzzle on the Continent in weaning a calf, and is occasionally fastened to the pole of a carriage to prevent the horses rubbing against it. From its fondness for insects the hedgehog is often placed in London kitchens and bakehouses to keep down the swarms of cockroaches with which they are infested.

EDENTATA.—There are a few of the toothless animals which are laid under contribution for food. Wallace tells us he found the flesh of the sloth tender and palatable. The echidna or native porcupine of Australia, which belongs to this order, is said to taste like a sucking pig. The flesh of the great scaly ant-eater (*Myrmecophaga jubata*) is esteemed a delicacy by the Indians and negroes in Brazil and Western Africa, and though black and of a strong musky flavour, is sometimes even met with at the tables of Europeans. Most of the varieties of armadillo are used for food in South America. The animal is roasted in its skin, and is esteemed one of the greatest delicacies of the country.

The flesh of the TATOO (*Dasypus* 12-*cinctus* and *D. villosus*) of South America is much esteemed. That of the hybrid species of the River Plate is one of the most exquisite meats that can be eaten. The animals have a scaly carapace or a hard shell-like armour which, divided into regular compartments, covers their head and body and often their tail.

THE PORCUPINE OR SPINY ANT-EATER (*Echidna hystrix*) is similar in its size and general appearance to the English hedge-hog, except that it has a long and slender snout, and has the power of protruding its tongue to a considerable distance. It is a burrowing animal, and although ordinarily of dull and slothful habits it makes its way into the ground with extraordinary rapidity; in fact, considering its size, its muscular power is very great. The upper surface of the body is covered with thick spines of a dirty yellowish colour, blackish at the points, and averaging about an inch and three quarters in length. Below these the body is covered with dark fur. When attacked or alarmed the animal will coil himself like a hedge-hog, and burying his nose in the earth, leave nothing but a round of prickles exposed, which neither man nor beast can touch with impunity. The genus Echidna has a very extensive distribution in Australia, but is generally found on the rocky coast, never far in the interior. These animals, of which there are two species, the brush-tailed and the branded ant-eater (*Myrmecobius fasciatus*), are not very numerous, but may frequently be met with even in the vicinity of Sydney by those acquainted with their habits and with the mode of tracing them. Feeding largely on ants' eggs, the flesh is delicate meat, resembling that of a young sucking pig, and considered superior to hare.

Of the pangolins or scaly ant-eaters there are several species.

Manis javanica is met with in Siam and the mountains and southern provinces of China. The dark yellow scales are much used medicinally in China for all sorts of maladies either calcined or in their natural state, being considered a remedy for ulcers and a cure for the itch, to which the Chinese are much subject. The scale is often fixed on a bamboo to scratch the skin by those troubled with the itch.

We now reach the Rodents or gnawing animals, and find many of these are of great service to man.

THE CANADIAN PORCUPINE (*Hystrix pilosus*, Catesby; *Erethrizon dorsatus*, Linn.).—The Indians and hunters in the United States and about the Rocky Mountain ranges eat the flesh of this animal, but to a more refined taste it would be unpalatable. The quills or spines are much used for embroidering the only really tasteful articles to be found among the natives of those regions and of which considerable quantities are sold in the American cities. The Slave Indians, dwelling along the McKenzie and Liard's Rivers, are reckoned the most skilful fabricators of this manufacture. The things made out of them consist of belts, bands, garters, bracelets; and they are also used for ornamenting birch bark-work, baskets, shot pouches, dresses, and shoes. In manufacturing belts, &c., a frame-work of sinew-thread is first laid, through which the quills are interwoven in squares, something in the manner of Berlin-wool work. The articles, when finished, are very pretty, and some of the women are sufficiently adepts to follow any angular pattern which may be set them.

The flesh of the common crested porcupine (*Hystrix cristata*) is considered very delicate food, and is often eaten at dinners in Rome, being sold at 5*d.* per pound, the porcupine being not uncommon in the Campagna. It is said they should be cooked like a hare or with wine sauce like a wild boar. It is hunted in Algeria, and the quills are exported from the ports of northern Africa; in 1873 about 21,000 valued at £60, were shipped from Morocco.

The flesh of the AGOUTI (*Dasyprocta*) of South America is firm, white, tender and well tasted, and when fat and well dressed is by no means unpalatable food. It has been sometimes termed the rabbit of South America. That of the CAVY (*D. acuchi*), a smaller species, also resembles it. The flesh of the GUINEA-PIG (*Cavia aperoea*) is white and savoury. That of the wild species common in Central America is very delicate.

CHINCHILLA LANIGERA.

The PACA or spotted cavy (*Cœlogenys fulgis* and *C. subniger*) is one of the best game animals of Brazil and breeds freely when domesticated. Its flesh is said to be very savoury and forms a staple article of food in many parts of South America.

The skins of the BIZCACHA or VIZCACHA (*Calomys bizcacha*) of

Buenos Ayres have occasionally found a market in England on account of their fur. The flesh of this animal when cooked is very white and good, but is seldom used, other animal food being so abundant in South America.

The fur of the CHINCHILLA (*Chinchilla lanigera*) a South American rodent, is remarkably soft, and is extensively used both in America and Europe. The skins reach us from Chili, Peru, New Granada, and Buenos Ayres. In 1837, 37,337 dozen of these skins were shipped from Buenos Ayres, valued at 16*s.* the dozen, and 2742 dozen from Arica. There is a specimen of the skin among the collection of furs in the Museum.

BEAVER (*Castor Canadensis*, Kuhl.). Beaver skins were at one time a very important article of commerce from America. The Hudson's Bay Company in 1743 sent to London 150,000 skins, and in 1808, no less than 127,000 were sent from Quebec alone to this country. In 1827 from more than four times the extent of territory, the amount did not exceed 50,000. But in the last quarter of a century they have again become greatly in demand. The imports by the Hudson's Bay Company have been as follows of late years:—

Year	Skins	Year	Skins
1856	72,482	1866	150,776
1857	80,640	1867	144,744
1858	90,217	1868	148,040
1859	107,428	1869	157,415
1860	91,459	1870	122,985
1861	106,270	1871	169,876
1862	116,144	1872	165,031
1863	114,442	1873	149,045
1864	143,238	1874	117,489
1865	118,575	1875	146,434

A beaver skin will be found in the collection of furs of the Museum, and there is also a fine stuffed specimen of a beaver caught in England.

In commerce beaver skins, cut open, stretched to a hoop and dried in the ordinary manner, are named "beaver parchment," and form by far the greatest part of the importation. When the beaver skins have been made into dresses, and worn by the Indians it is termed "beaver coat," and though it may have been in use a whole season, it still brings a good price. Inferior sized skins are named "beaver cub."

The beaver exists some distance within the arctic circle, and the darkest coloured pelts come from Fort Good Hope. The Slave and Dog-rib tribes make capotes and robes out of the skin.

BEAVER (*Castor fiber*).

Beavers are caught in good order at all seasons of the year in the Rocky Mountains, as there it is never warm enough to injure the fur; in the low lands, however, along the Missouri, the trappers rarely commence their hunting before September, and relinquish it about the last of May. Sixty or seventy skins make a pack of 100 lbs.

Beaver hair was formerly much employed with the fur of hares and rabbits for making hats, but these heavy beaver hats have gone out of fashion. Cloth was also made of it, but it was heavy, dear and not agreeable in appearance.

The flesh and tail of the beaver are amongst the most prized dainties of Indian epicures. It used to be considered best when roasted in the skin after the hair had been singed off, and in some districts it required all the influence of the fur traders to restrain the hunters from sacrificing a considerable quantity of beaver fur every year to secure the enjoyment of this luxury. The flesh resembles pork in its flavour, but Sir John Richardson tells us that it requires a strong stomach to sustain a full meal of it.

Castoreum.—The beaver besides its fur furnishes a substance used in medicine as an antispasmodic, and commercially known as castoreum. The principal imports are by the Hudson's Bay Company, and these vary from 1,000 to 5,000 lbs. annually. In 1875, 3297 lbs. were received. The price ranges from 7*s.* to 25*s.* per lb. The taste of Siberian castoreum is much more pronounced than that of Canadian in consequence of its greater richness in castorine, of which it contains 4·6 per cent, whilst Canadian contains but 1·98 per cent.

In the hunting districts castoreum is extensively used for enticing the lynx to enter into the snaring cabins.

The castoreum in its recent state has an orange colour, which deepens as it dries, into bright reddish-brown. During the drying, which is allowed to go on in the shade, a gummy matter exudes through the sac, which the Indians delight in eating. It is never adulterated in the fur countries.

In the trade district of Alaska in the seventy-six years ending with 1862, 2,500 lbs of castoreum was obtained.

MUSK RAT or MUSQUASH. (*Fiber Zibethicus*). The geographical range of this animal is extensive, from 30° to 60° north.

Though they have a strong musky flavour, particularly in spring, their flesh is eaten by the Indians in North America, who prize it for a time when it is fat, but soon tire of it. The musky odour is owing to a whitish fluid deposited in certain glands near the origin of the tail. The skin, when taken from the body, still retains the scent. The fur on the whole body is soft and glossy, and beneath is a finer fur or thick down as in the beaver, but shorter and less lustrous.

Still it is a beautiful fur, and when dyed and plucked. its resemblance to the fur seal is so great as to deceive any but dealers. Twenty years ago they used to fetch a shilling a skin.

MUSK RAT (*Fiber Zibethicus*).

Great numbers of these skins were formerly used in America, and sent to Europe for making beaver hats; but since the general use of silk hats, the demand for this purpose has passed away. They are however now in great request as a fur. In 1824 only about 5000 skins were exported to Great Britain from America; now the quantities collected by the Hudson's Bay Company are enormous, ranging between 400,000 and 600,000 annually. The lowest number sent of late years has been 177,000.

The European musk rat is met with about the River Volga, and the adjacent lakes from Novgorod to Saratov. The price at Orenburg in olden times used to be as low as 7*d.* a hundred for these skins. And even in later years they were so common near Nishni Novgorod that the peasants would bring in 500 each to market, which they sold at 3*s.* 6*d.* a hundred. The skins and tails were put into chests and wardrobes in Russia to preserve clothes from moths. The skins were also believed to guard the wearers of them from fevers and pestilence.

HAMSTER (*Cricetus vulgare*, F. Cuv.; *Mus cricetus*, Linn.).—The fur of this animal is useful for many purposes, and they are sometimes found entirely black. Hamsters are great depredators to corn, and hence ruthlessly destroyed. In some years as many as 80,000 skins of these little animals have been brought to the town-hall of Gotha.

The skins of the Berwitski or Siberian mice (*Mus striatus*, Linn.) are used as a fur.

The Bohemian ladies used to make cloaks of the skins of the Souslik or Zizel—(*Mus citellus*, Linn.). The spotted kinds make very beautiful linings.

THE BROWN RAT (*Mus decumanus*, Linn.). This very destructive animal, according to the accounts of historians, came from Asia to Europe, about the beginning of the seventeenth century; was unknown in England before 1730, and, according to Dr. Harlan, did not make its appearance in North America until the year 1775. Rat skins are said to be used in Paris for making gloves.

MARMOTS.—There are several marmots in the northern parts of America, viz., the hoary marmot, *Arctomys pruinosus*, Gmelin, *A. kennicotii*, and *A. monax.* The latter is known as the woodchuck in the United States. Out of the skins of all these the mountain tribes of Indians make robes, and the flesh is considered good when fat.

THE QUEBEC MARMOT (*Arctomys empetra*, Schreber), inhabits the woody districts from Canada to lat. 61° and perhaps still further north. Its fur is not of much value. The hoary marmot, or whistler (*A. pruinosus*, Pennant), is of the size of the woodchuck or Maryland marmot. It is covered with a beautiful long silver-gray hair and has a long bushy tail. The Indians consider their flesh as delicious food; by sewing a number of their skins together they make good blankets. Some of the spines are entirely white, others brown at the tips.

QUEBEC MARMOT (*Arctomys empetra*).

Ground squirrel or Sifleur. The Tawny marmot (*Arctomys Richardsonii*). The fur is stout and fine, and of two kinds—a short down and longer and coarser hairs. 3639 marmot skins were imported from North America in 1860, some are received under the name of Weenusk. They are obtained from the *Arctomys empetra*, *Spermophilus Parryi*, Rich.

The European marmots are :—

1. The Polish marmot (*A. bobac*).
2. The Marmot of the Alps (*A. marmotta*).
3. Souslik, or marmot of the Volga (*A.* [*Spermophilus*] *guttatus*).
4. Zizil, Susel or Hungarian marmot (*A. citellus vel undulatus*).
5. Siberian marmot or jevraschka (*A. concolor*) and two or three others. These do not produce so fine and close a covering as the fur of the North American squirrels.

From 60,000 to 120,000 marmots are caught every year in the country about Aschersleben in Middle Germany, when it retires to its winter quarters; in its fat condition, the marmot is killed in great numbers and eaten in parts of Europe.

Under the name of NUTRIA, a very large number of the skins of the *Myopotamus coypus* have been in some years imported from South America. This rodent has two kinds of fur, long ruddy hair, which gives the tone of colour, and a brownish ash-coloured fur at its base, which, like that of the beaver, used to be largely employed in the hat manufacture. The habits of the coypou are much like those of the other aquatic rodent animals.

In 1837 we imported from Buenos Ayres 622,236 skins, valued at sixpence each; in 1841 and 1842 about 1,200,000 were imported yearly; in 1850 only 18,713 were received; in 1858, 856,130, valued at about 9½*d.* each; and in 1861, 506,089.

RABBITS.—The Rabbit (*Lepus cuniculus*), originally a native of Spain, belongs to the same genus as the hare, but differs from it considerably in its gregarious habits, its subterranean mode of life, its whiter flesh, and the less perfect state of the young when first produced.

"Fancy" rabbits, as they are called, have been kept in this country from the earliest times, being mentioned by Cæsar in his account of Britain. They have been raised from the common rabbit by scientific breeding in many countries, particularly in Persia and Arabia, whence we have derived several of our present breeds. The silver-grey rabbit is much bred for its fur, which fetches a high price in China, and is sometimes used here.

In France about 60,000,000 of rabbit skins are collected annually. The principal portion are used for their hair, for hat making, and a few millions for furs, when dyed different shades. More than sixteen different transformations are given to the skins in France, and they are made to assume the appearance of costly furs to suit the tastes and customs of different nations. France works up seven-eighths of the quantity, and Belgium the other eighth.

The number of rabbits sold in the markets of Paris in 1845 was only 177,000, but in 1860 this number had increased to more than two millions. Taking the number consumed in Paris at a thirtieth of the whole quantity used in the Republic, the total consumption would be about 70 millions annually. According to official returns, the average price of a rabbit exceeds two francs, so that the value of those consumed annually may be taken at £5,500,000.

Messrs. Leeding & Co., Borough High Street, state that they work up yearly nearly two million hare and rabbit skins, nearly all English, for foreign skins do not make such good fur as the English and Scotch. They buy the skins principally from poulterers, marine store dealers, and hawkers. They estimate the number of skins manufactured in this country annually produced and imported at not less than 30,000,000. At an average weight of 2½ pounds each, these would give 35,500 tons of food, and valuing each at but 1*s*., this amounts to £1,500,000, to which £500,000 more must be added for the value of the skins. The number of rabbits sold by licensed game dealers in the kingdom in 1872 was returned at about 5,200,000, but these are a very small proportion of the total sales.

It was stated by Lord Malmesbury in the House of Lords in March, 1873, that Mr. Brooks, the salesman, in Leadenhall Market, receives weekly from Ostend 1500 cases, each containing 100 rabbits; that one gentleman in Somerset sends to Birmingham 10,000 a week, to Wolverhampton 3000, and to Nottingham 4,000 weekly. Southampton receives in the year about 90,000.

The skins of many species of hares and rabbits are valuable for common purposes of fur, on account of the almost inexhaustible supply. The most common American species are the *Lepus americanus*, *L. palustris*, *L. sylvaticus*, *L. townsendii*, *L. glacialis*, *L. aquaticus*, *L. nigricaudatus*. The colours vary from light grey to yellowish and reddish brown in summer; in winter white predominates. The fur of the polar hare (*L. glacialis*) is beautifully white and soft, and is sometimes substituted for ermine. When beaver hats were worn the felt bodies were made of rabbit skins;

it is now dyed and made into a great variety of common articles. Rabbits' hair and wool has been made in England and America into a kind of cloth for ladies' wear, mixed with wool and cotton. There is a specimen in the Museum made from the Angora rabbit. When properly prepared the hair affords a good strong yarn, which is said to be in no way inferior to wool. An important industry is likely to grow out of the successful introduction of rabbit hair weaving in all countries. It is rather remarkable that the use of the hair has not been thought of before, particularly when we consider how many hundred millions of rabbits are annually destroyed. No animal is better adapted for raising on a large scale than the rabbit. They multiply almost as rapidly as white mice, and are not confined to any particular climate.

In one of the Jury Reports of the French International Exhibition of 1867, after speaking of the choice varieties of Angora and Cashmere rabbits, raised for the fine hair of their fur, often an inch to an inch and a half long, which is spun into silky yarn in parts of France, and which is so much sought after in North America for making certain shawls that £3 are often given for fur skins of rabbits only five or six months old, it goes on to speak of the profit to be derived from raising rabbits. The Ministry of Public Assistance, Paris, buys rabbit skins regularly at 6*d.* per lb. Rabbits breed so rapidly, that they will have from 6 to 8 litters annually, of from 8 to 12 and 16 young ones.

The wild rabbit's skin is more valuable than that of the tame, and is largely used in the manufacture of hats. Some of the cheaper kinds of muffs, boas, &c., are made of hare and rabbit skins, which are dyed to imitate the skins of foreign animals. There are about eighty fur-cutting machines in London and the provinces, each cutting four hundred and twenty skins a week. Each machine employs about twenty-five workpeople on the premises, or two thousand all over the kingdom, earning nearly £60,000 per annum in weekly wages. About eight thousand persons are also employed out of doors in collecting, sorting, drying, and otherwise preparing the skins during four months of the year, and their earnings amount to something like £140,000 during the season. Thus rabbits, as furnishing a most appreciable quantity of animal food, and contributing to several manufacturing industries, are by no means to be despised.

In the Museum collection of furs will be found skins of the wild rabbit, the blue rabbit, dyed rabbit skin, Polish rabbit, silver rabbit, and the English hare, while in the Waste Products Collection are further illustrations of the economic uses of rabbit and hare skins and the wool.

In all the Australian colonies rabbits have multiplied prodigiously, and the difficulty is to keep down their increase. In South Australia the destruction is carried on to such an extent that in one district alone in four months capitations were paid on 26,000 head. Formerly payment was made on the production of the rabbits' tails, but it was discovered that the men were guilty of the deception and cruelty of cutting the tails off and letting the rabbits go alive. They are now obliged to kill them, and produce the ears and scalps before they get paid.

In Tasmania rabbits are so numerous that in the Campbell Town, Ross, and other districts, it is not unusual for a shooting party of three to kill several hundred in an afternoon. On some farms the bodies of the rabbits are boiled down as food for the pigs. In other parts the skins are taken off and sent to market, or are used by the men on the farms for rugs and bedding.

During the year 1872, 218 bales containing 48,238 dozen of Tasmanian rabbit skins were exported, valued at £5,860. From the 1st of January 1872 up to the first quarter of 1873, there were over 700,000 rabbit skins shipped, valued there at £7,252.

Rabbits have increased also in Victoria to such an extent since their introduction, that men have been specially employed to destroy them. The rapid growth of this comparatively trivial industry of rabbit skins, illustrates the commercial activity of the colony. When first they were an article of trade their value was

6d. a dozen, whereas they are now worth *2s.* in Melbourne, and sell in London at *3s. 6d.* a dozen. About 100,000 dozen are shipped annually. The Australians have also begun to preserve rabbits in tins, stewed, curried, and fricasseed, for export to England.

Hares.—The number of hares consumed in the United Kingdom is very large. Those sold by licensed game dealers in 1872 were returned at 702,830; but this by no means represents the whole supply. It may be computed that 150,000 at least are sold annually in the London markets and shops. The white hare comes to us from the north of Scotland, where it changes colour, like those in the Arctic regions.

In France the skin of the hare is not used as a fur. In this country they were employed to a small extent as chest protectors, but other and cheaper substitutes have now been introduced by the chemists.

In the fur countries the American hare becomes white in the winter. This change takes place in the northern districts in the month of October, and the animals retain their white coat until the end of April, when it begins to fall off, and is replaced by their shorter and coloured summer dress. The winter skins of this animal are imported by the Hudson's Bay Company under the name of "rabbit skins," as it is much like that of the common European rabbit. In some parts of the fur countries the natives line their dresses with hare skins, and the Hare Indians sometimes tear the skins with the fur into strips, and plait them into a kind of cloth. They resort to this expedient, however, only from the scarcity of deer skins and moose leather, which form closer and better dresses.

The Polar Hare.—The winter fur of this animal (*Lepus glacialis*, Leach) is of a snow-white colour to the roots, and is more dense, and of a finer quality than that of the American hare. It bears a close resemblance to swandown. The weight of a full-grown Polar hare varies, according to its condition, from

7 to 14 lbs., and a similar variation in the weight of the common British hare is known to exist. Its flesh is whitish and well flavoured, being greatly superior to that of the American hare (*L. Americanus*, Erx.); and also much more juicy than the alpine or varying hare of Scotland.

The Polar hare inhabits both sides of Baffin's Bay, and is common on the barren grounds at the northern extremity of the American continent. Its southern boundary is about the elevated ridge of the Rocky Mountains and the eastern coast of Labrador. The fur is in prime order in lat. 65° about the end of October, and begins towards the end of April to be replaced by the summer coat, which is more or less coloured.

THE AMERICAN RABBIT (*L. campestris*, Bach), so essential to the welfare of the Chippewyan nation, is spread all over the district, except upon the barren grounds. It is subject to periodical failures, which occur with great regularity, and cause no small amount of privation and suffering to the Indians, when they happen. When the animals are numerous, the Tinné tribes of the McKenzie valley subsist altogether on them, and the skins furnish almost entirely their winter clothing—robes, shirts, capotes, mittens, and socks being made, which afford a sufficient protection against the most severe cold, though they do not form lasting garments, as the hair falls out very quickly.

The import of rabbit skins by the Hudson's Bay Company varies, having been as low as 18,500 skins in 1862, and as high as 141,400 in 1866.

SQUIRRELS are found in various countries of the world, but they vary in size and colour. Those of temperate Europe have usually a reddish fur on the back, and white beneath, but occasionally this deepens to a dark brown; in Russia and other northern parts, and in Siberia, the fur becomes of a slate or grey colour in winter, and a large number of the skins enter into commerce. Many millions are obtained in the Russian territories, and a good many are sent to England. The fur is much used as linings for tippets and cuffs, for which its softness and cheapness peculiarly adapt it. So numerous are the squirrels in parts of California, that a bill has passed the Legislature which requires every land-owner to exterminate them on his land, and in case of his failure to do so, authorizes a public officer to attend to the matter, the expense to be a lien on his land, to be collected in the same manner as taxes. The skins of the ground squirrel, or hackee, (*Tamias Lysteri*), have been tanned. They are found remarkably strong, and as well suited for gloves as the very best kid, which the skin somewhat resembles, but is much finer.

The North American red squirrel (*Sciurus hudsonius*, Pennant). The fur was at one time of no value, but is now an article of trade. The flesh is tender and edible, but that of the male has a strong murine flavour. The other species most esteemed are, *S. carolinensis*, cat squirrel (*S. cinereus*), black (*S. niger*), grey (*S. migratorius*), and fox squirrel (*S. capistratus*).

The grey squirrel was long a favourite fur with the Chinese, and held the chief place in the barter trade at Kiachta. At the end of the last century 7,000,000 of these skins were exchanged, and the export rose in 1800 to 10,000,000 skins; but from that period it gradually declined, till between 1849 and 1852 the average annual export fell to 1,460,000. In 1853 the quantity dropped to 344,557. In 1858 it rose to about 1,100,000, and in a year or two ceased altogether. The fur of the Russian squirrel was formerly largely used also in Europe, but is now seldom seen.

The grey squirrel and black squirrel make excellent pies. The flesh tastes like that of a rabbit, but much more juicy. The grey squirrel is a common dish in Virginia. It is usually broiled, and is very palatable. The large squirrel of the East is excellent eating, and its flesh is much valued in Borneo.

In the fur collection of the Museum will be found specimens of the red, black, blue, grey, and Indian squirrels, and a stuffed specimen of the flying squirrel (*Pteromys volucella*).

CHAPTER XI.

SOLIDUNGULA AND PACHYDERMS.

In this chapter we revert again to the larger and useful animals, especially to the thick-skinned ones. The numbers and uses of the horse, ass, and mule in various countries are described, and the economic products of horse hides and horse hair, with their applications noticed. The African and Asiatic elephant, and the extinct mammoth elephant receive notice, and very complete statistics of the ivory trade are given.

THE HORSE, the ASS and the MULE are the useful servants of man in various countries, and the services they render cannot be overrated, for draught and for the saddle, as well as for the secondary products obtained from them after death.

It would be quite impossible in a limited space to go into detail as to the several breeds of horses, and therefore a very brief mention of one or two must suffice.

Dray and Waggon Horses.—Ordinary sized farm horses weigh from 12 to 13 cwt.; riding or harness horses from 10 to 11 cwt. Amongst the heavy weights on record are a horse which belonged to the Carron Co. that weighed 18¼ cwt. and one to Barclay, Perkins & Co., the brewers, which weighed one ton.

A kind of horse still in demand is what is termed the dray horse, thus described in the *Farmer's Magazine.*

There are two very important points in these—namely, size, and a disposition to accumulate fat; and this latter property is as essential as it is in cattle or sheep. To breed horses of gigantic proportions, the first impression that naturally arises with the uninitiated, is that of procuring mares of great size and selecting for them partners of still greater magnitude. There is nothing, however, more uncertain in this respect than the produce of very

X 2

large mares. It not unfrequently occurs that a medium sized—and sometimes indeed, a small sized—mare breeds very large foals, and in the event of one of such offspring being employed as a brood mare in the hope that her progeny will be equal to, or perhaps exceed, the proportion of the dam, that she gives birth to mere insignificant foals. It often happens that her produce will in this respect follow their grandam, such is the propensity of nature to go back to originals. Very large mares will frequently produce one or two foals of great size and many of insignificant proportions.

The heavy-heeled, lethargic mare of elephantine proportions, is far from being likely to produce foals that will realize great prices as dray-horses in London, but one with a roomy frame, a great body on short clean legs, strong shoulders, and vast development in the loins and quarters, put to a stallion of adequate proportions, will probably produce what is required, but with the utmost caution, success is by no means certain.

There is another class of horses in great request in London, commanding high prices, of lighter make, the breeding of which is less precarious ; and they are annually becoming more and more in demand, while the dray-horse is becoming less in vogue. These are employed for working luggage-vans, in connection with the railway stations, and for other purposes of heavy draught in which great power is required, and, to meet the usage of the times, more activity and pace than the heavy dray or common cart horses possess. To produce this description of horse a good-shaped active cart mare or a powerful mare of lighter make, is crossed with a sire of the Yorkshire or Cleveland breed.

They must have action and be able to trot at the rate of seven or eight miles an hour. Their height should not be less than sixteen hands.

For farming operations, in which they may be advantageously employed prior to the time when they have attained an age to command the top price of the London markets, they are incalculably superior to the huge lumbering dray-horses, which for agricultural purposes, conducted as they are at this time, are unwieldy, slow and useless brutes. They consume an enormous quantity of food, without which it is futile to expect that their gigantic frames can attain their proper growth; and to make them fat for sale, the indispensable property they must possess, involves an outlay that leaves but an insignificant margin for profit.

The Arab Horse.—The horses bred in Africa, in Egypt, in Arabia, in Syria, in Asiatic Turkey, in Persia, &c., are all of the same family, constituting the race known as of eastern blood. Strength, agility, vigorousness in form as in action, these are the characteristics of the horse the moment he is found beyond the Euphrates or below the Mediterranean and the Caucasus. Where it is on the land of Islamism, it is always temperate, invincible to privation and fatigue, living between sand and sky.

Whether it be known under the names of Turkish, Persian, Numidian, Barb or Arab of Syria matters little, these are mere prefixes and the name of the family is the horse of the East. The other variety, above the Mediterranean, is the European horse.

So thoroughly established is the excellence of the desert horse as compared with the more domesticated kinds of the Arab horse, that in selecting Arabs for turf purposes, it is thought among the first essentials to pick one that has been all his life in the desert.

The Arabs have a laconic statement that there are essentials in a horse which may be thus enumerated:

Four things large: the front or brow, the breast, the rump, the limbs. Four things long: the chest, the upper ribs, the belly, the haunches. Four things short: the back, the pasterns, the ears, the tail.

To this Abd-el-Kader added, that three things should be clear or pure:—the eyes, the skin, and the hoofs.

Statistics of Horses.—From the summary we have been able to make there would appear to be, in round numbers, about 39,000,000 horses in Europe, 16,500,000 in America and the West Indies, nearly 1,000,000 in Australasia and the Pacific Islands, and 500,000 in South Africa. For the great continent of Asia, there are no statistics available, nor as to the numbers of the equine race in Mexico, and the five Central American States, British Guiana, Brazil, Bolivia, Peru, Chili and Paraguay. About $7\frac{1}{2}$ million mules and asses have to be added to this number.

HORSES in various countries according to the latest returns.

Country	Year	Horses
EUROPE.		
Russia	1876,	21,570,000
Sweden	1874,	455,907
Norway	1865,	150,000
Denmark	1874,	350,000
Iceland	1875,	29,000
German Empire	1873,	3,352,231
Holland	1873,	253,394
Belgium	1866,	283,163
France	1874,	3,633,605
Portugal	1870,	79,716
Spain	1865,	672,559
Italy	1874,	500,000
Austria Proper	1871,	1,367,023
Hungary	1870,	2,179,811
Switzerland	1868,	100,000
Greece	1867,	98,938
Ionian Isles	1860,	12,000
Turkey	1874,	1,100,000
Great Britain	1875,	2,255,129
Ireland	1875,	534,883
AFRICA.		
Egypt	1871,	19,359
Algeria	1861,	165,402
Cape Colony	1875,	207,318
Orange Free State	1859,	27,552
Natal	1874,	23,443
ASIATIC ISLANDS.		
Java	1873,	632,164
Réunion	1874,	4,155
Mauritius	1875,	18,969
Ceylon	1874,	7,960
WEST INDIES.		
Jamaica	1869,	74,157
Martinique and Guadaloupe	1874,	9,502
Barbados	(estimated),	5,000
Other West India Islands	(estimated),	10,000
AMERICA.		
United States	1875,	9,504,200
Argentine Confederation	1875,	4,000,000
Uruguay	1873,	1,600,000
Canadian Dominion	1871,	2,624,290
Other Brit. Amer. Provinces	1871,	100,470
AUSTRALASIA.		
New South Wales	1875,	346,691
Queensland	1874,	99,243
Victoria	1875,	180,254
South Australia	1875,	93,122
Western Australia	1875,	26,636
Tasmania	1875,	23,473
New Zealand	1874,	99,261
Hawaiian Isles	1866,	25,000

Formerly, the only means for travelling from place to place was on the back of a horse; and then, as might be expected, a breed of horses with adequate powers to carry men was studiously kept up. But improvements in roads, and the construction of wheel carriages, have long afforded a different and more easy mode of conveyance. Besides, we now have railways and locomotive steam-power whisking the traveller (ay, and his horse too) along at a marvellous rate, annihilating space, to some station or terminus, whence he takes refuge in the mob of a 'bus, or in a cab, drawn by miserable horses, equal only to the short distance they have to drag the load.

In the year 1832, before railways were generally formed, the total number of horses used for riding and drawing carriages in Great Britain which contributed to the assessed taxes was 182,878; in 1862 their number was 306,798, besides 263,391 used in trade, &c.; and in 1872 they had increased to 907,000. The duty has now been abolished.

The nominal horse-power of the engineer—the steam-horse—is so rapidly taking the place of the *bonâ fide* draught of the equine animal that we are beginning to overlook, in many instances, our old friend the horse, and yet he does not decline in price or demand. The 7000 locomotives running over some 16,000 miles of rail in the United Kingdom annually, and those traversing a greater mileage in the United States, to say nothing of the continent and the colonies, and South America, are doing much of the haulage in pleasure and traffic for which we had formerly to depend upon horses. Still, to mount the large armies kept up, to move the artillery, ambulances, &c., and to facilitate the traffic of towns and railway stations, horses continue in great request, although their opponent—steam—is fast driving them from the plough-team and other farm uses. The difficulty to obtain horses of suitable powers as remounts for cavalry is yearly increasing.

War makes sad havoc among horses. The civil war in the United States reduced the number of horses in that great republic fully one million. The average destruction in the Federal army alone was not less than 500 per day. The Franco-German war also entailed a heavy destruction of horses.

The scarcity of horses in England is becoming a matter of general anxiety, not only to individuals who require the use of horses, but to the Government, which has to make provision for the service of the cavalry and artillery. Lord Rosebery, in a debate in the House of Lords three years ago, stated that while we now had 6,600 cavalry horses and 6,000 artillery horses, in case of war we should require 2,500 cavalry horses and 4,000 artillery horses more, and not fewer than 25,000 light and 75,000 heavy transport horses.

We have been largely exporting horses of late years, which if continued must make us shortly dependent on a foreign supply. We exported in

	No.	Value.
1872	12,618	£348,620
1873	17,822	585,868
1874	12,039	535,771
1875	25,757	980,777

Russia is essentially the land of horses. Compared with other countries in Europe we find that Russia has one horse for every three and a half persons, Austria, one for every ten, Prussia and Great Britain, one for every eleven, France, one for every twelve, and Italy, one for every twenty-seven. In Russia the greater number are to be found in the provinces of Oran and Perm (where most of the inhabitants, who are of the Tartar race, have a peculiar inclination for horse breeding), in the country of the Don Cossacks, where horsemanship is an indispensable part of the daily avocations of the people, and in the provinces of Middle Russia, which require a great number of horses to carry on their extensive trade.

As far back as the historical accounts of Russia extend, the rearing of horses seems always to have formed a notable branch of the national industry. The warlike and nomadic habits of the ancient population—the increasing demands for the supply of an extensive army—the immense distances, requiring a large amount of animal labour, as well for the conveyance of produce and merchandise as for locomotion; all these combined have stimulated the development of this branch of rural economy, favoured as it is, over a large portion of the empire by the great extent of pasture lands. Accordingly we find the Russians possess excellent horses for all uses, and a larger number than all the other European States combined.

The government, true to the traditions of the country, continues to devote special attention to the encouragement of horse-breeding, for which the steppes of south-eastern Russia afford the most favourable opportunity.

There are no less than twelve imperial studs, of which nine are in Europe and three in the Caucasus. There are also fourteen government depôts and country stables spread over the empire, each having from 60 to 150 stallions for public use. There are also about 2,500 private studs, which own 6,500 stallions and 69,000 brood mares. The government encourages by every means in its power the improvement in the breeds by sustaining racehorses of pure blood and trotting horses, by offering prizes and medals at races and exhibitions. There are about 400 horse fairs held annually, and at fifteen of these from 2,000 to 10,000 horses are offered for sale, and some 300,000 change hands. The price of horses has advanced of late years in Russia fully 50 to 100 per cent.

The export of horses from Russia is shown by the following figures, giving the average number in the last three quinquennial periods.

	Total.	Annual Average.
1861 to 1865	51,000	10,000
1866 to 1870	65,000	13,000
1871 to 1875	125,000	25,000

Taking the mean value fixed in the Russian official documents £12, this gives a present annual total of £300,000.

In the provinces of Central Asia and Siberia there is almost a horse, to every inhabitant, the proportion being 179.3 to 100 in the former, 70·1 to 100 in the latter. Next to these stand the eastern and south-eastern governments. The third place is occupied by the central governments of European Russia, where horse breeding by great landholders, as well as the production of cart-horses, is carried on most extensively.

Russia requires, in time of war, for its army purposes, 272,000 horses; Italy, 26,000; Turkey, the same number; Egypt has about 6,000 mounted troops; Austria, 59,000 cavalry, and as many more horses are required for the engineer and artillery service; China has 30,000 Mongolian cavalry; Brazil, 3,000 cavalry.

Spain stands third in rank among the European States for the character of its equine race. Andalusia is the principal district for horses. The colts of Cordova are renowned, so are those of Galicia and Valencia, while New Castile is famous for its mules, and Estramadura for its asses.

The Indian Government have made the Cape Colony a remount station, and have purchased horses largely of late years there. The native horse of the Colony would not satisfy the eye of an English remount officer, but his endurance during the longest and hottest day surpasses that of almost any other horse. The horse is small; is generally brought for inspection in very bad condition, and requires a thoroughly honest and good judge of horses to do justice to the breeders and dealers, and to protect the Government. There are now more than 200,000 horses in the Cape Colony.

In the Eastern countries beyond Bengal the horse is seldom used for draught, and the race is a diminutive one, the ponies being only employed for riding now and then. He is but poorly fitted for carrying burdens, and useless for the plough, the cart, or cavalry purposes. Java, however, possesses a race of very hardy ponies, useful for riding or draught, which numbers over 600,000.

The island of Sumbawa is noted for the number and beauty of its small horses, the most esteemed of all the Indian Archipelago, and these are largely exported to Java.

While horses have increased in number in New South Wales

from 116,397 in 1850 to 346,000 in 1875, they have deteriorated in quality. Breeding does not pay because the rules of good breeding are neglected so much, that the reputation of Australian horses has failed to attract buyers. Were the breed good the demand would soon increase more for shipment to India.

There are splendid horses in the country, but the tendency to breed from any sort of mare leads to the production of a wholly worthless class of horses, which are commonly sold in the markets for a few shillings a head, or boiled down for the grease to lubricate the carriages they should draw.

Wild horses, or "brumbies," as they are termed, are so numerous in some of the districts that persons are employed exclusively to shoot them, who receive a fixed sum besides the hide and hair as perquisites. Twenty years ago a settler was offered £10 a head all round for the horses on his run, which he refused, and now he has to pay for shooting them. At one station alone, in the district of Goondiwindi, Queensland, over 7,000 wild horses have been shot, and yet plenty still remain.

The Australian horse, although not so large as the English horse, is active, muscular, and hardy to a degree. The results obtained, even under the present system, are such as to prove that the climate and soil of Australia are in the highest degree favourable to the breeding and rearing of the horse; and when the time arrives that the breeder can fence extensively without fear of removal, it will be seen that Australia can and will produce the finest horses in the world. The Indian Government, as well as civilians in India, require a supply of horses every year, of the very best description the wilds of Australia produce; not a large, overgrown animal, but a stout, compact horse of about 15 hands —a wild horse, the result of promiscuous breeding. Such are the wild horses of the South American pampas, which are among the most enduring in the world; and such are the wild horses of Australia that are now depasturing there in mobs of thousands. The drive to the port of shipment tames them, and they are easily broken in, and when landed are fit for every purpose required in India. Horses admirably adapted for the Indian market are becoming so numerous in the up-country districts, whilst the demand for them is so limited, and the prices are so low, that it is found more profitable to shoot them for the hair and hides, than to send them to the city for sale. And many station-holders view the presence of horses upon their runs as an absolute nuisance. The Indian market offers a panacea for the ills complained of by breeders of horse stock, and would probably return to them a handsome remuneration if the trade were pursued with spirit. About 2,600 horses are imported annually into India chiefly from Australia.

After Russia, the United States and the South American republics have the largest number of horses. They are so numerous in Cordoba and other parts of the State of Buenos Ayres, that droves of mares are annually exported to Peru; the breed, especially in the sierras, is small.

Thirty years ago, all travelling in the country was done on horse-back, and although there are now railways traversing the country, the natives make light of galloping 100 or even 150 miles in a day. When we find that mares are sold to the slaughter-houses at a dollar a head, to be killed merely for their hide and grease, the rearing of an improved breed for export could not but prove remunerative.

In France it is made a subject of regret that the ancient native breeds have disappeared, and the reproach of Buffon holds good, "that they have augmented the number and confusion by favouring the mixture of races, and without reflection have interfered with nature, introducing African and Asiatic horses, so that the primitive races of France have been lost, and their characteristics are scarcely to be distinguished. These characters would be" (he observes) "more specially marked, and the differences be more sensible, if the races of each climate were kept separate, and unchanged by inter-breeding."

In France, including Algeria, the number of horses was, by the latest returns, 3,630,000.

When the peace establishment of the army required 70,000 saddle horses, France had to import annually from 6,000 to 7,000; that establishment is now raised to 90,000. To pass from a peace to a war footing under the new military law demands 176,000 horses, and a large proportion of them of superior quality to what was formerly considered serviceable. There were in 1874 107,000 horses in the army, and gendarmerie, and republican guard. The movements of field artillery are now so much accelerated that many of the draught horses of that arm must come from the same class as those for the cavalry.

Great Britain requires about 40,000 horses for her military service, and 5,000 for her Irish mounted police.

In Germany, for military purposes, 97,000 horses are required on the peace footing, and on a war footing 233,500 for field troops, 30,500 for depot troops, and 37,000 for garrison purposes. Her cavalry requires about 70,000.

Having glanced at the value of the horse when living, let us now consider its commercial products after death. Of late years, owing to the high price of butcher's meat on the Continent, there has been a large consumption of horse-flesh for human food.

After this noble animal has worn out all its powerful energies in serving man, it leaves him a carcase from which much can be utilised for manufactures and industry.*

Horse Hides are valuable when tanned, and considerable numbers are obtained in Russia and the River Plate States. It has already been stated at page 155 that 1,000,000 horse hides are tanned annually in Russia. The Mongols and Kirghiz make articles of dress of horse leather. Several Tartar tribes are experts in leather plaiting, and use a great deal of horse leather for this purpose.

The following figures will give an idea of the export trade in horse hides from Buenos Ayres :—

Year	Number	Year	Number
1853	129,905	1872	208,509
1860	278,613	1873	149,482
1870	102,250		

In 1870 about 19,000 cwts. of horse hides of the value of 30,000*l.* of British produce were exported. The supply of foreign horse hides to this country varies, as will be seen by the following figures :—

Year	Number	Year	Number
1853	171,300	1869	98,821
1854	206,500	1870	101,940
1862	160,186	1871	151,123

The current prices in the London market at the close of 1876 for horse hides were, for salted hides 10*s.* to 15*s.* each; for dry hides 7*s.* to 8*s.*; tanned hides, Spanish, weighing 6 lbs. to 12 lbs. 16*d.* to 24*d.* per lb.; English hides of 10 lbs. to 18 lbs., 13*d.* to 16*d.* Horse butts about the same.

Horse Hair.—Though in nearly every country horse hair is collected, the chief sources of supply are Germany, the River

* In the cheap descriptive Catalogue of the Waste Products Collection in the Museum, the statistics of the consumption of horse flesh as food, and the relative value of the several parts of the dead horse are fully described, as well as in my work on "Waste Substances," published by Hardwicke and Bogue.

Plate States, Belgium, and Russia, as the following list of the imports into the United Kingdom in 1875 will show :—

	cwts.
Germany	7,299
Uruguay	4,935
Argentine Republic	3,069
Belgium	1,556
Australia	906
Brazil	996
Russia	708
Other Countries	762
	20,231

We get some good long tail hair from Russia; other qualities, long, medium, and short, from South America, where a quantity is obtained from the immense number of horses, which, in a wild state, roam over the pampas of that continent. The average annual shipments of horse hair from Rio Janeiro in the five years ending 1873 were 10,714 cwts. and from the River Plate States 9,000 cwts. The manes and tails of horses which die in this country, form but a small portion of the supply; the hair thus obtained is generally of poor quality, and unfit for use in the manufacture of hair cloth. The foreign material is imported in bales weighing about 1,000 lbs. each. These contain either "mixed hair," that is, hair of different lengths, or else are filled entirely with long hair.

As the material in its raw state is in a tangled and dirty condition, the first process through which it passes is sorting, during which the different coloured hairs are placed in separate heaps. This is to facilitate the subsequent dyeing, as the black stain used is much more readily imparted to hair that is naturally of a dark colour than to that of the lighter shades. The bundles of sorted hair are then hackled, by which process the hairs are made straight, and the foreign substances and dirt mingled with them removed. During the hackling, great care is taken not to break the hair, as upon its length its value depends—long hair being much more scarce than the shorter varieties, and consequently far more costly. A number of tufts of hackled hair are next placed between the teeth of a couple of cards, or flat pieces of tough wood, on which pointed spikes of steel of about three inches in length are inserted. One of these cards is placed on its back on a table and the bundles of hair laid side by side between its teeth; when this card is full, the other one is placed upon it, points down, so that the bundles are firmly held by the double set of spikes. The hair it must be remembered, is still of different lengths, and it is the object of this carding to arrange the long and short hairs in separate bundles. The workman, therefore, begins by pulling out from the bundles between the cards, all of the long hair in the ends nearest to him, and then keeps on removing more and more, until the set of extremities at which he is working are perfectly even, no one hair projecting more than another. Then he fastens the ends, removes the upper card, reverses the bundles, and repeats the same process with the other extremities. When he finishes, the hairs between the cards are all of exactly the same length, and the separate tufts are now ready either to be made into curled hair, to be sold to the brush-makers, or to be woven into hair cloth.

From the bales it is sometimes thrown into a "picker" making 800 revolutions per minute, and then twisted into ropes by machinery, to make it curl. The next process is to boil it, that it may be thoroughly cleansed, for which purpose it is put into vats, heated with exhausted steam from the engine; this done, it is thoroughly dried in an oven. The ropes of hair are then ready to be picked into pieces for use.

The process of curling is begun by making the lengths of hair which are found to be too short for other uses, into a rope. The workman, taking a bundle of loose material in his hand, attaches it to a revolving hook, and, walking backwards, continually adding more hair, spins a long, tight strand. Two of these strands are twisted into a cord, which, when finished, is reeled up into large coils. It is then boiled and immediately afterwards baked, this process setting the "kink" in the hairs, and rendering them thoroughly elastic. In this condition, curled hair is sold to the trade; it only remains to untwist and pick out the rope by hand to obtain the desired quantities.

The short hair is serviceable after curling, for stuffing chair seats, cushions, sofas, mattresses, &c.; the long hair for weaving into seating and covering; and the middle lengths for brush-making in lieu of bristles. Light horse hair can be dyed of various colours; but as there is only a limited supply of the pure white, some difficulty would arise in obtaining sufficient raw material.

Hair-cloth, principally used for covering sofas and chairs, is manufactured from the longer and better qualities of hair. The bundles of hair destined to be made into cloth are removed to the dye-house. There they are attached to a large iron grating, which, when filled, is lowered into a vat of boiling dye, in which it remains for about five hours. The hair is then detached, and is ready for weaving. The warp of the cloth is of black cotton thread. Linen thread is a better material, but makes a stiffer and harsher fabric, less suited for upholsterers' uses. The hair composes the weft, and its length depends upon the width of cloth to be made, the usual proportion being a thirty-five inch hair to a thirty-inch cloth.

Hair-seating used to be woven by hand, every hair being introduced singly. It differs in this respect from most other woven fabrics, in which there is a uniform and continuous supply of material.

In the Animal Collection, there are some interesting specimens of damask hair-cloths.* Among these are fancy green-striped hair seating, plain grey satin seating, orange damask figured, scarlet

* Case **93** contains horse hair of different kinds and qualities, horse hair cloth used for bags, sacks and rope made of it, and combed hair for stuffing. Case **94**, an instructive collection of horse hair, bleached and dyed different colours, for making hair-cloth seatings; also a series showing the process of making horse hair gloves and flesh rubbers.

damask, figured black diaper damask, plain black satin hair, &c. In some of these specimens a variety of damask patterns or designs are introduced by the application of the Jacquard loom, and also a diversity of colours.

After leaving the loom, the cloth is pressed between hot metal plates, and then rubbed to give it the necessary polish. As furnished to the trade it is generally black, and its principal use is, as already mentioned, for covering furniture. A very fine variety is sometimes made for sieves. In price, hair-cloth averages about 4*s.* per yard. In width, it is manufactured in all sizes between fourteen and thirty-two inches.

Horse hair-cloth originated in England, and in the early part of the century M. Bardel received medals for the introduction of the manufacture into France. The manufacture is now largely carried on in the United States, where upwards of 20,000,000 yards are made yearly; one factory alone makes 600,000 yards annually, no hairs less than 16 inches long answer for the purpose. The invention of special looms which pick-up and place the hairs automatically, has driven out of the market the hand-made goods.

Contrary to the popular idea, the hair is not, as a rule, round. A section under the microscope shows a form as though a third of a circle had been cut off, and the flat portion slightly indented. This conformation caused some difficulties in the manipulation, which required great skill and the most delicate machinery to overcome.

Among the various other purposes to which horse hair is applied, are, for making ladies stiff petticoats, mixed with cotton, for bags for pressing apples, cloth for straining purposes used by brewers, oil refiners, &c., for socks or soles for lining boots and shoes.

Horse hair is used on the helmets of the Horse Guards. It is made into ropes and wigs,—even the learned lawyer is obliged to rob the poor Siberian horse for his wig. Horse hair shirts were formerly worn for the health of the soul, but gloves of the same material are now used for the health of the body. False tails of horse hair are made for the use of those horses which are deficient in that respect. Fishing lines are occasionally made of horse hair.

A queue, or tail of horse hair, suspended at the end of a pike, terminated by a gilded pennant, is the Turkish standard, or emblem of authority. Commanders are distinguished by the number of horse tails carried before them, or planted in front of their tents. Thus the Sultan has seven, the Grand Vizier five, and the Pashas three, two, or one. The usage of these tails is of Tartaric origin.

The imports of horse hair into the United Kindom are shown in the following return:

	Cwts.	Value.		Cwts.	Value.
1862	17,478	£86,199	1869	19,857	£152,631
1863	15,854	78,813	1870	14,134	127,145
1864	17,743	90,511	1871	21,365	202,103
1865	21,078	106,938	1872	15,830	188,726
1866	20,875	113,097	1873	19,891	223,393
1867	17,318	100,336	1874	19,102	177,782
1868	16,056	102,509	1875	20,231	171,053

MULES are the well-known offspring of the ass and the mare. From being very hardy and sure footed they are much employed in the West Indies and South America. Peru is noted for the finest breed of mules in America. The plains of Caraccas are estimated to furnish annually 30,000 for the West India islands. There are no precise data as to the numbers, but there are probably from one to two millions in the different South American States.

In the two provinces of Catamarco and Tucuman there are more than 57,000 mules and asses. At the Atajo mines in the former there are 3,500 employed on Mr. Lafone's mines, besides large numbers at other mines.

Mules and asses, especially the former, are not very abundant in the United Kingdom. Asses are employed in drawing small carts and in carrying burdens, by hucksters in town and by poor people in the country, particularly those near commons, the barrenest of which will keep an ass. But it is chiefly in South America and the West Indies where mules are used. The importance of mules in carrying on the traffic of Brazil is evidenced by the fact stated some years ago by a well-informed writer, that the arrivals of laden mules at Cubatas, near the port of Santos, in the province of San Paulo, amounted to 420,000. The introduction of railroads in Brazil is doing something to replace them, but it is only transferring their services to other quarters. The mules annually imported from the cattle farms of Sorocaba in San Paulo, into the province of Minas, average 60,000 to 70,000 a year. Minas itself rears a considerable number of mules; those in actual service there may be rated at about 260,000; the number that die there annually are from 18,000 to 20,000. Mules constitute the great commerce of Tucuman. These animals are bought in Cordova, Sante Fé, and Buenos Ayres, and being fattened during the winter in the valleys and plains of the States of the Argentine Confederation, are driven to Peru, where about 50,000 are sold annually. Many are taken thence to Truxillo, Lima, and other places both on the coast and in the interior, for sale. In some of the North American States numbers are raised at a better profit than horses, from the fact that they come to maturity much earlier and command remunerating prices at any age.

In 1783, Washington became convinced of the defective nature of the working animals employed in the Southern States, and set about remedying the evil by the introduction of mules instead of horses, the mules being found to live longer, be less liable to disease, require less food, and in every respect to be more valuable and economical than the horse in agricultural labour in the South.

In Spain, horses are comparatively little used compared with mules and asses. According to the census of 1865 there were in the Peninsula and the two adjacent provinces about 1,000,000 mules and 1,300,000 asses, compared with 627,000 horses. The mules of the South of Europe are frequently sixteen or seventeen hands high.

Notwithstanding the improved means of transport introduced into the country, many troops of mules for carrying burdens are kept up in Spain for supplying the towns of the interior. The Iberians of Andalusia, the Minglanilla in New Castile, the inhabitants of Segovia and of Maragalirie in Old Castile, and many others besides, preserve the old system of transport by mules' backs. Various causes contribute to keep up this system. The taste of families for the occupation of muleteers, in which they have been brought up, and which has been handed down from father to son; the high rates of tolls on roads which the muleteers can avoid by short cuts and circuitous routes; the heavy rates charged for goods by railway, and the risk and damage on the transport of delicate and fragile goods which are tossed about at stations, contribute to keep up the old mode. Railways instead of destroying the old traffic, have rather augmented it.

The mules of Morocco excel even those of Spain. Some of these, Dr. Leared states, are as large as a full-sized horse. Plodding, patient, sure-footed and docile, they carry the traveller at the rate of four miles an hour through a long day, with few halts and little sustenance. The mule in Morocco holds the place of the stage coach, or railway train in more advanced countries. The price of a mule ranges there between £20 and £40. Mules are beginning to be much used in Central Asia. There has of late years been a development of the through traffic from the Punjab to Turkestan by mules. A troop of 170 went from Jalandhar to Ladak, which at once obtained return freights to carry borax to Kulu. They are found to carry heavier weights than the camel, and to travel faster.

Asses.—Persia is famous for its races of asses, of which there are three principal varieties, one white and two grey, of different sizes.

The former are the most esteemed. The two larger varieties attain the height of four and a half feet, and are much finer than the wild ass. The smaller grey ass is a very pretty animal from the south of Persia. The skin of the ass was formerly used for making shagreen and other leather.

Statistics of MULES and ASSES:—

Russia . . .	1874,	30,000	Brazil . .	(estimate)	400,000
Spain . . .	1865,	2,292,692	Argentine Republic	1875,	235,000
Portugal . .	1860,	200,000	Uruguay . .	1872,	120,000
France .	1872,	749,754	Algeria . . .	1861,	310,831
Austria . . .	1860,	200,000	Morocco .	(estimate)	1,000,000
Germany .	(estimate)	20,000	Cape Colony . .	1875,	29,517
Holland . . .	1864,	2,696	Martinique and		
Ireland . .	1875,	180,000	Guadaloupe .	1874,	7,264
United States .	1874,	1,500,000	Réunion . .	1874,	9,295

The QUAGGA (*Equus Quagga*) has been tamed. The Hottentots and other natives of South Africa hunt the animal for its flesh. Lieut. Moodie, in his "Ten Years in South Africa," says he once had a piece of the flesh cooked for his breakfast, grilled and peppered, and he did not find it at all unpalatable. It was, he adds, certainly better than the horse-flesh which was served to him in the hospital at Bergen-op-Zoom in 1814. Captain Burton, in his "Central Africa," tells us that of wild flesh the favourite is that of the Zebra (*E. Zebra*); it is smoked or jerked, despite which it retains a most savoury flavour.

The Roman peasants found the flesh of the ass palatable, and the celebrated Mæcenas having tasted it, introduced it to the tables of the great and rich, but the fashion of eating it lasted no longer than his life. Galen compares the flesh of the ass to that of the stag. It is said to be eaten plentifully in the guinguettes of Paris, under the denomination of veal.

A NATIVE ACCIDENTALLY BORNE AWAY BY A QUAGGA.

PACHYDERMATA.

THE ELEPHANT.—There are at present but two existing species of this huge pachyderm, and its principal economic product is the ivory which it furnishes to commerce.

Besides these contemporary races of elephants, the market is extensively supplied with the ivory derived from the tusks of the great mammoth or fossil elephant of the geologist. The remains of this gigantic animal are abundantly distributed over the whole extent of the globe. They exist in large masses in the northern hemisphere, deeply embedded in the alluvial deposits of the tertiary period. Humboldt discovered specimens on some of the most elevated ridges of the Andes; and similar remains have been found in Africa. In the frozen regions of the far North, surrounded by successive layers of everlasting ice, the fossil ivory exists in a state of perfect preservation, and it constitutes an important article of commerce in the north of Europe.

A true elephant roamed in countless herds over the temperate and northern parts of Europe, Asia, and America. This was the creature called by the Russians, mammoth; it was warmly clad with both hair and fur, as became an animal deriving its sustenance from the leaves and branches of trees which grow as high as the 65th degree of north latitude.

Formerly the name ivory was given to the main substance of the teeth of all animals; but it is now, by the best anatomists and physiologists, restricted to that modification of dentine or tooth-substance which, in transverse sections or fractures, shows lines of different colours or striæ proceeding in the arc of a circle, and forming by their decussations minute curvilinear lozenge-shaped spaces. By this character, which is presented in even the smallest portion of an elephant's tusk in transverse section or fracture, true ivory may be distinguished from every other kind of tooth-substance, and from every counterfeit, whether derived from tooth or bone. This engine-turned decussating appearance is as characteristic of fossil as of recent ivory. Although, however, no other teeth except those of the elephant present this characteristic of true ivory, there are teeth in many other species of animals which, from their large size and the density of their principal substance,

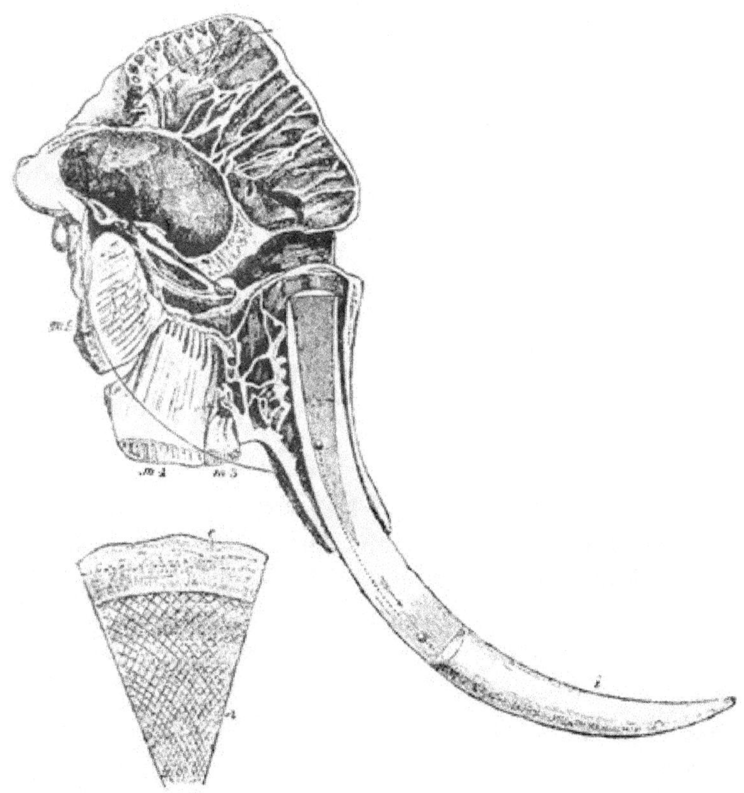

SECTION OF SKULL OF AN ELEPHANT, SHOWING THE DENTITION AND A BULLET LODGED IN IT.

are useful in the arts for purposes analogous to those for which true ivory is used; and some of these dental tissues, such as those of the large tusks of the hippopotamus, are more serviceable for certain purposes than any other kind of tooth-substance.

The dentition of the genus *Elephas*, includes two long tusks (see *i* in the annexed figure), one in each of the premaxillary bones, and large and complex molars (*m* 3, *m* 5, *m* 4) in both jaws: of the latter there is never more than one wholly, or two partially, in place and use on each side at any given time, the series continually being in progress of formation and destruction, of shedding and replacement; and, in the elephants, all the grinders succeed one another, horizontally, from behind forwards.

In a lecture by Prof. Owen delivered before the Society of Arts, in Dec., 1856, "On the Ivory and Teeth of Commerce," from which I quote, he remarks that the utility of teeth in commerce and the arts depends chiefly on a peculiar modification in their laws of growth. For the most part, teeth—as in our own frames—having attained a certain size and shape, cease to grow. They are incapable of renewing the waste to which they are liable through daily use, and when worn away or affected by decay, they perish. Certain teeth in man and the majority of the mammalia, are succeeded by a new tooth after the old one is worn away and shed, but this with very few exceptions, occurs but once in the course of life. Teeth of this kind are said to be of limited growth, but there are other teeth, such as the front teeth of the rat, rabbit, and all the rodent tribe, the tusks of the boar and hippopotamus, the long descending canine tusks of the walrus, the still longer spiral horn-like tusk of the narwhal, and the ivory tusks of the elephant, which are endowed with the property of perpetual growth, that is, they grow as long as the animal lives.

The molar teeth of the elephant are of little value. They have been attempted to be utilised for making knife handles, but owing to the larger proportion of cement to dentine they are brittle: small slabs are sometimes sliced out of them. It will be seen by the appended illustrations that there is a striking difference in the enamel plates of the molars of the African and Asiatic elephants: in the former the lamellar divisions of the crown are fewer and thicker, and they expand more uniformly from the margin to the centre, yielding a lozenge-form when cut or worn transversely, as in mastication.

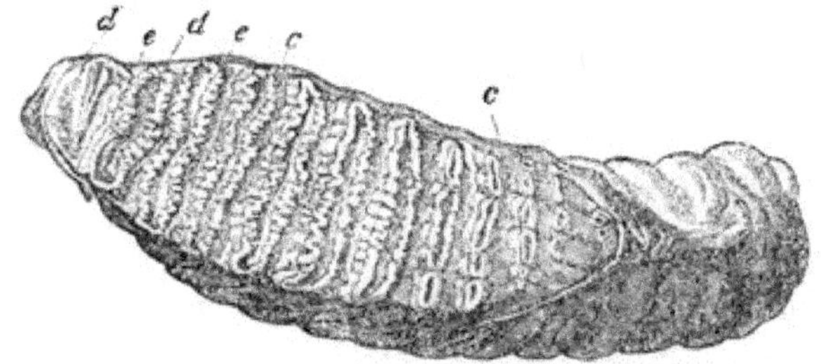

UPPER MOLAR TOOTH OF INDIAN ELEPHANT (*Elephas Indicus*), SHOWING TRANSVERSE ARRANGEMENT OF THE ENAMEL PLATES (*dd*), AND FESTOONING OF THEIR BORDERS (*ee*).

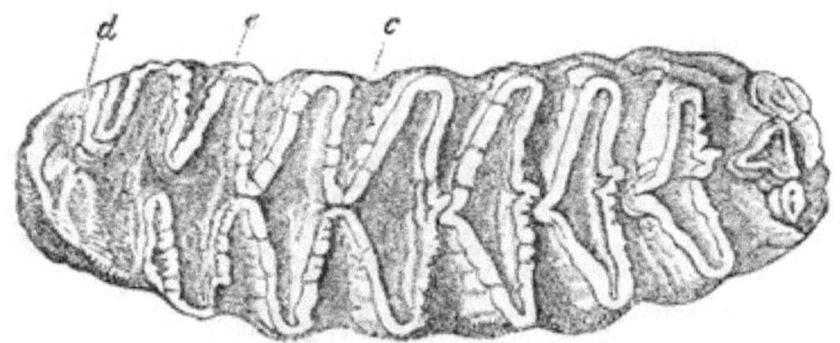

UPPER MOLAR TOOTH OF AFRICAN ELEPHANT (*Elephas Africanus*), SHOWING THE LOZENGE-SHAPED ARRANGEMENT OF THE ENAMEL PLATES OR RIDGES.

The AFRICAN ELEPHANT is much larger than that of Asia; sixteen feet is not an uncommon height, and many are killed of even superior proportions.

The distinction between the African and Indian elephants is clearly defined by Dr. Livingstone. "The first elephant (he says) killed by my man was a male, not full grown; his height at the withers was 8 feet 4 inches; and the circumference of his fore foot 44 inches. The female, which was full grown, measured in height 8 feet 8 inches, the circumference of the fore foot being 48 inches. These details are given with the view of showing that the general rule, that twice the circumference of the fore foot equals the height of the animal is not of universal application; for in the first instance, double the circumference falls short of the height by twelve inches, and in the second instance by eight inches. Subsequent observations, however, proved the general correctness of the rule with regard to full-grown animals.

"The greater size of the African elephant in the south would at once distinguish it from the Indian one; but here they approach more nearly to each other in bulk, a female being about as large as a common Indian male. But the ear of the African is an external mark which no one will mistake even in a picture. That of the female now killed was 4 feet 5 inches in depth, and 4 feet in horizontal breadth, and I have seen a native creep under one so as to be quite covered by it. The ear of the Indian variety is not more than a third of this size." Livingstone adds that although there was more abundant food he found the animals of all kinds in the districts north of 20°, smaller than the same races existing southward of that latitude.

The full-grown male elephants on the river Zouga, seemed no larger than the females on the Lempopo, while they were even smaller than on the Zouga. There is, however, an increase in the size of the tusks as we approach the equator.

It would seem from old Roman imperial medals extant, that in early centuries of the Christian era the African elephant was tamed; although no attempt has been made to subjugate this most useful animal at the Cape.

African Ivory.—The finest transparent ivory is collected principally along the western coast of Africa, within 10° north and south of the equator, and it is considered to become worse in quality and more liable to be found damaged with the increase of latitude in either direction. The best white ivory is for the most part the produce of the eastern coast of Africa generally. The African ivory in the best condition, should appear, when recently cut, of a mellow, warm, transparent tint, almost as if soaked in oil, and with little appearance of grain or fibre. In this state it is

AFRICAN ELEPHANTS.

termed "transparent" or "green" ivory. But on exposure, the transparency to a certain extent goes off, and the ivory is left of a delicate white hue, which should be permanently retained.

Among large African tusks may be enumerated a pair weighing 325 lbs., each measuring 8 ft. 6 in. in length, and 22 inches in basal circumference; and one weighing 110 lbs. from South Africa, which was shown by Mr. Joseph Cawood, of Graham's Town, at the Great Exhibition in 1851; one weighing 139 lbs., shown by Messrs. Fauntleroy, and one weighing 103 lbs., shown by Messrs. Buchanan and Law at the same Exhibition. Some

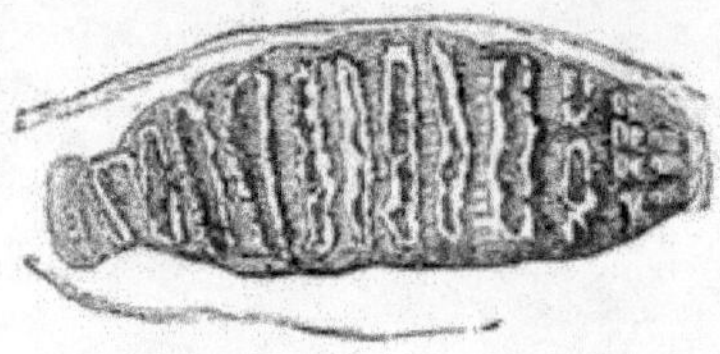

very large tusks were also shown at the London Exhibition in 1862. Mr. Gordon Cumming had one weighing 173 lbs. A tusk was shipped from Camaroon to Liverpool, some years ago, weighing 164 lbs.; and another to Bristol weighing 147 lbs. Captain R. Burton, writing on the ivory-trade of Zanzibar, states that the tusks are larger there than elsewhere. At Mozambique, for instance, 60 lbs. would be a good average for a lot. At Zanzibar, a lot of 47 averaged 95 lbs.; 80 lbs. is considered a moderate average, and 70 to 75 lbs. poor. Monster tusks are spoken of. Specimens of 175 lbs. are not very rare, and the people have traditions that these wonderful armatures have extended to 227 lbs., and even to 280 lbs. each, approximating in size to the huge fossil tusks occasionally met with in the northern regions. Cuvier made a list of the largest tusks found up to his time, and the most considerable one registered by him weighed 350 lbs. At a sale of tusks in London, the largest brought from Bombay and Zanzibar weighed 122 lbs. Those from Angola averaged 69 lbs., from the Cape and Natal 106 lbs., from Lagos

114 lbs., and from Gaboon 91 lbs. But these are by no means the largest sizes to be found at present; for elephant-hunters now penetrate further into Africa, and therefore meet with older animals.

In former years, before European commerce had extended far into the interior of Africa, many of the native chiefs surrounded their dwellings with palisades of elephants' tusks, but now that their intrinsic value has become well known, they have been bartered away for more appreciated objects. Some of the natives often spoil their tusks by cutting them. The Gold Coast ivory may generally be known by having a rough-hewn hole made near the end of the hollow. Some tribes stain the exterior by sticking the tooth in the sooty rafters of their chimneyless huts, with the idea that so treated it will not crack or split in the sun.

The following statement will serve to show the localities from whence our supplies of African ivory have been received of late years, given in cwts.

	1870.	1871.	1872.	1873.	1874.	1875.
Portuguese Possessions	410	417	395	449	387	312
West Africa, not designated ...	1322	1966	2081	2465	1958	2297
Gold Coast	...	...	131	78	102	251
South Africa	1185	1109	1019	1117	1548	1442
East Coast	...	823	450	69	303	696
Egypt	4521	5242	4932	4628	2297	2411
Tripoli and Tunis	...	...	...	...	141	400

Some African ivory also comes in through Malta and Portugal.

The quality of the African ivory varies considerably. That best adapted for the English market comes from the Camaroon coast, and the most esteemed runs from 50 lbs. weight upwards, the next in value 35 lbs. to 50 lbs., then from 18 lbs. to 35 lbs. Tusks are valued in proportion to their size; those that weigh 1 cwt. or more are the best, and fetch from £40 to £50 the cwt.; the second class comprehends such as require two teeth to make a cwt. or more; the third class, three or more to the cwt. All below 18 lbs. are called "scrivelloes," and are of the least value, except such as are adapted for cutting billiard balls.

Gaboon, Loando, Congo, and Ambriz, rank next to the Camaroons in the quality of the ivory shipped.

The Gold Coast ivory, and that brought down for shipment to Sierra Leone and Cape Coast Castle, ranks next, and is tolerably good. Gambia teeth are usually very bad, always broken, very crooked, cracked in the hollows, and more or less damaged.

The Gold Coast ivory may generally be known by having a rough hewn hole made near the end of the hollow, and these teeth (the term under which they pass in commerce) are much esteemed.

Fine ivory is known by having no cracks or flaws, either in the solid or in the hollow. Cracks in the ivory are a serious detriment, and must be always particularly noticed. The elephants' tusks that are only rather tapering in shape are most liked; very crooked teeth must be guarded against, as they cut up to great disadvantage. Broken-pointed teeth, or those with deep flaws, or otherwise damaged about the point, must be avoided. Teeth with large hollows are not at all liked, as there must inevitably be a great waste in cutting them up; in short, a fine tooth is known by being of a neat tapering shape, with but a small hollow, free from cracks, with a fine, thin, clear coat, free from flaws; it is also transparent, which may be discovered by holding the point to a candle.

Quantity and value of the ivory exported from Natal:

	lbs.	£		lbs.	£
1850	49,481	9,200	1863	183,014	40,736
1851	40,917	7,989	1864	125,874	26,254
1852	31,542	6,274	1865	79,395	19,154
1853	25,672	8,634	1866	29,218	6,673
1854	64,209	14,686	1867	21,618	5,908
1855	55,567	13,504	1868	34,000	9,169
1856	58,450	13,716	1869	39,590	11,326
1857	68,273	18,170	1870	50,038	14,350
1858	117,663	31,754	1871	51,597	13,220
1859	67,255	17,618	1872	36,052	9,392
1860	73,695	21,064	1873	48,986	17,199
1861	75,545	22,825	1874	27,678	9,036
1862	10,116	27,059			

—*Stat. Abstract, Colonial Possessions.*

Quantities and values of exports of ivory from the Cape Colony:

	lbs.	£		lbs.	£
1857	57,757	16,358	1866	40,969	10,114
1858	92,944	21,903	1867	61,514	11,018
1859	30,362	8,398	1868	36,669	8,772
1860	63,323	16,826	1869	55,155	13,482
1861	58,330	14,731	1870	60,419	14,546
1862	113,379	24,813	1871	46,382	11,566
1863	53,226	10,773	1872	88,203	24,176
1864	26,013	4,488	1873	91,457	32,613
1865	30,241	10,486	1874	73,747	26,667

THE ASIATIC ELEPHANT.—The elephant is still found in the east and north of the island of Ceylon in very large numbers. A good many are killed annually by sportsmen. The late Major Rogers, of sporting notoriety, is known to have killed in that island upwards of 1,000 with his own rifle; 1,600 were captured by being snared, or enclosed in kraals, and exported to India from the northern province in the five years ending 1862. When tamed and used as a beast of burden, it is quite surprising what he can be made to do, and the intelligence and docility he displays—his cunning, strength, and endurance are well known. He is often employed by the Public Works Department on the roads and bridges, and is found a valuable and frequently indispensable auxiliary. He will roll over large stones, place them in position as directed, arrange them in twos with the skill of a mason, or level a beam with the eye of a builder. In matters of mere brute strength they find him almost equally useful. They have large waggons made expressly for him. When suitably harnessed he is made to draw immense loads. The cut on the next page represents an elephant at work in a timber yard at Moulmein.

Elephants exist in large troops in the forests of Siam, and are caught by the aid of domesticated females, who entice them into ambushes or snares prepared beforehand. They are soon tamed, and when domesticated render great services to their owners. The ivory of Siam is much sought after for its quality and its density, being considered superior to that obtained from parts of India. A fine tusk of a Siamese elephant was shown at the London Exhibition of 1851 that weighed 100 lbs., the ivory of which was very white and compact.

INDIAN ELEPHANT EMPLOYED IN A TIMBER YARD, MOULMEIN.

In some "Notes on Northern Cachar," by Lieut. R. Stewart in a number of the Bengal Asiatic Society some information is given as to the capture of elephants by the natives:—

"Elephants are slain in great numbers by the Kookies wherever they are to be had, not only the tusks but the flesh being highly prized. Parties of twenty and upwards go out in pursuit of them at a time. When some recent elephant track is discovered in the forest, two or three of the party ascend some convenient tree, whose branches overhang the track, the remainder follow it up, and having got on the other side of the herd, scare it towards the ambush by shouting, beating gongs, and discharging fire-arms. Here, while passing, the animals are assailed from above with long spears having huge iron bars covered with deadly poison; every wound inflicted results in the death of the animal at not more than half a mile from the spot on which he was hit. So wary are the elephants, however, that it is seldom more than two out of a herd are killed. At the place where their game is found dead they commence cutting him up, and extract his tusks; laden with these and as much of the flesh as they can carry, they return home, and other parties go out and encamp in the neighbourhood of the carcase until they have entirely consumed it or are driven away by the effluvia of decomposition. Portions of the flesh that they cannot immediately eat, are dried and smoked to be kept for future consumption."

The ivory from India is of a more opaque or dead white, but has a tendency to turn yellow or discoloured. The teeth from Ceylon and Singapore are distinguished by their fine grain, pearly blush colour, and small size.

In the Asiatic elephants, tusks of a size which make them valuable as ivory in commerce are peculiar to the males, while in the African elephants, both males and females afford good size tusks. Tusks weighing 150 lbs. have been occasionally exported from Pegu and Cochin China.

Several specimens of fine ivory were exhibited at the local show Madras in 1854. The largest tusks sent were a pair weighing 138 pounds, obtained from a wild elephant killed in the Travancore forests. One tusk weighed 71 pounds, the other 67 pounds, and showed a fine white compact kind of ivory; of these two, one measured six feet eight inches and the other six feet six inches, the circumference being seventeen inches in each case.

It is a singular fact that the domestication of the elephant is usually attended by deterioration of the length and quality of the ivory.

Quantity and value of imports of Asiatic ivory into the United Kingdom from British India and Aden:

	Cwts.	£		Cwts.	£
1849	4,058	—	1863	2,520	70,013
1850	3,440	—	1864	2,620	82,549
1851	2,580	—	1865	2,058	65,913
1852	4,425	—	1866	1,515	56,855
1853	3,123	—	1867	493	17,814
1854	3,166	80,733	1868	80	2,741
1855	2,748	68,874	1869	132	4,717
1856	4,781	170,523	1870	221	7,933
1857	3,232	138,653	1871	399	11,966
1858	5,096	170,860	1872	244	8,059
1859	2,254	70,409	1873	1,201	58,082
1860	2,849	87,118	1874	3,942	193,299
1861	4,339	116,188	1875	3,560	171,227
1862	3,428	78,010			

Value of the ivory and ivory-ware exported from India for a series of years:

	£		£		£
1849	70,828	1858	19,805	1867	85,008
1850	56,718	1859	98,157	1868	64,575
1851	43,086	1860	97,126	1869	122,520
1852	90,140	1861	33,039	1870	108,289
1853	55,886	1862	120,367	1871	83,003
1854	80,895	1863	60,260	1872	61,918
1855	66,921	1864	80,398	1873	104,869
1856	82,384	1865	77,217	1874	127,404
1857	128,096	1866	92,402	1875	89,640

In the last named year the exports were from Bengal £1,843, from British Burmah £2,231, and the remainder from Madras and Sind.

As illustrations there are in the Animal Products Collection a fine small elephant's skull with tusks; nine large ivory tusks, one weighing 94 lbs., and many small ones, samples of mammoth ivory, and diseased ivory, ivory anklets and numerous manufacturing applications of ivory, with elephants' bones from Siam.

The Ivory Trade.—The supply of ivory seems to increase rather than to diminish. In 1827 we only received 3,000 cwt. Taking the imports in round numbers, in 1840 they were nearly

SIAMESE WAR ELEPHANT.

5,500 cwt.; in 1850, 9,400 cwt.; in 1860 nearly 10,000 cwt.; in 1870, 12,600 and in 1875, 16,258 cwt. The total quantity of ivory imported into the United Kingdom in the quarter of a century ending with 1875 was 267,924 cwt. There is a small quantity of other ivory than elephants' tusks mixed up with this, but it is comparatively unimportant, and it would probably not amount to 8,000 cwt. in the whole period. Now averaging the tusks received at only 26 lbs. each, nearly 21,000 elephants must have been slaughtered annually, exclusive of what was shipped elsewhere, or used locally, and supposing some tusks were cast, some animals died a natural death, and some portion was fossil ivory, at least 20,000 elephants annually will be within the mark.

Quantities and value of the ivory imported into the United Kingdom:

	Cwt.	£		Cwt.	£
1840	5,469		1858	12,279	410,608
1841	5,811		1859	10,821	336,147
1842	6,429		1860	10,854	332,166
1843	5,462		1861	11,163	297,491
1844	5,229		1862	11,605	262,962
1845	6,402		1863	9,290	256,059
1846	5,394		1864	11,497	361,384
1847	6,477		1865	10,268	322,286
1848	5,146		1866	11,982	445,335
1849	7,457		1867	10,343	360,520
1850	9,396		1868	9,909	337,403
1851	6,356		1869	14,599	507,319
1852	9,515		1870	12,590	439,839
1853	10,388		1871	12,797	341,205
1854	9,299	230,420	1872	11,229	348,693
1855	8,376	219,964	1873	13,385	506,629
1856	9,866	343,517	1874	13,365	576,915
1857	9,890	421,318	1875	16,258	772,371

From the following statement of the exports of ivory, it will be seen that we retain and use up about half the quantity imported.

Quantities and value of the ivory re-exported from the United Kingdom:

	Cwt.	£		Cwt.	£
1868	5,927	202,901	1872	7,703	290,960
1869	5,837	207,505	1873	5,228	238,548
1870	4,039	144,192	1874	7,930	366,920
1871	7,519	196,005	1875	7,546	399,119

EXTINCT ELEPHANTS.—The tusks of the extinct *Elephas primigenius*, or mammoth, have a bolder and more extensive curvature than those of the *Elephas indicus*. Some have been found which describe a circle, but the curve being oblique they thus clear the head, and point outwards, downwards, and backwards. The numerous fossil tusks of the mammoth which have been discovered and recorded, may be ranged under two averages of size—the larger ones at nine feet and a-half, the smaller at five feet and a-half in length. Professor Owen, in his "History of British

HEAD OF INDIAN ELEPHANT.

Fossil Mammals," vol. viii. p. 247, has assigned reasons for the probability of the latter belonging to the female mammoth, which must accordingly have differed from the existing elephant of India, and have more resembled that of Africa in the development of her tusks; yet manifesting an intermediate character by their smaller size.

In the account of the mammoth bones and teeth of Siberia,

published in the "Philosophical Transactions" for 1737 (No. 446), tusks are cited which weighed 200 lbs. each, "used as ivory to make combs, boxes, and such other things; being a little more brittle, and easily turning yellow by weather and heat." From that time to the present, there has been little intermission in the supply of ivory furnished by the tusks of the extinct elephant of a former world.

This *fossil ivory* formed one of the earliest articles of export from Siberia to China. About 20,000 lbs. of it (that is to say, the tusks of at least 100 mammoth-elephants), are bartered every year in New Siberia, so that in a period of 200 years of trade with that country, the tusks of 20,000 mammoths must have been disposed of—perhaps even twice that number, since only 200 lbs. of ivory are calculated as the average weight produced by a pair of these tusks. As many as ten of these tusks have been found lying together in the Tundra, weighing from 150 to 300 lbs. each. The largest are rarely seen out of the country, many of them being too rotten to be made use of; while others are so large that they cannot be carried away, and are sawn up into blocks or slabs on the spot where they are found, with very considerable waste, so that the loss of weight in the produce of a tusk before the ivory comes to market is of no trifling amount. A large portion of this ivory is used by the nomad tribes in their sledges, arms, and household implements, and formerly a great quantity used to be exported to China; a trade which can be traced back to a very distant period, for Giovanni de Plano Caspini, a Franciscan monk sent by Pope Innocent IV. in 1246 into Tartary, describes a magnificent throne of carved ivory, richly ornamented with gold and precious stones, belonging to the Tartar Khan of the Golden Horde, the work of a Russian jeweller, the slabs of which were so large that they could only have been cut out of great mammoth tusks.

Notwithstanding the enormous amount already carried away, the stores of fossil-ivory do not appear to diminish; in many places, near the mouths of the great rivers flowing into the Arctic Ocean, the bones and tusks of these antediluvian pachyderms lie scattered about like the relics of a ploughed-up battle-field; while in other parts these creatures of a former world seem to have huddled together in herds for protection against the sudden destruction that befell them, since their remains are found lying together in heaps. In 1821, a hunter from Yakutsk, on the Lena, found in the New Siberia Islands alone, 500 poods (18,000 lbs.) of mammoth tusks, none of which weighed more than 3 poods; and this notwithstanding that another hunter on a previous visit in 1809, had brought away with him 250 poods of ivory from the same islands. The inhabitants on the mainland, pile up in heaps the tusks which are found scattered about on the Tundra, and convey them in large boats up the Lena. In the period from 1825 to 1831, at least 1,500 poods of 36 lbs. each reached Yakutsk yearly. The trade in fossil-ivory at Turuchansk, on the Jenissee, has for many years past amounted to from 80 to 100 poods annually, and that of Obidorsk, on the river Obi, to from 75 to 100 poods.

Entire mammoths have occasionally been discovered not only with the skin (which was protected with a double covering of hair and wool) entire, but with the fleshy portions of the body in such a state of preservation, that they have afforded food to dogs and wild beasts in the neighbourhood of the places where they were found. They appear to have been suddenly enveloped in ice, or to have sunk into mud when on the point of congealing, and which before the process of decay could commence, froze around the bodies, and has preserved them up to the present time in the condition in which they perished. It is thus they are occasionally found when a landslip occurs in the frozen soil of the Siberian coast, which never thaws, even during the greatest heat of the summer, to a depth of more than two feet, and in this way, within a period of a century and a half, five or six of these curious corpses have come to light from their icy graves.

A very perfect specimen of the mammoth in this state was

discovered in the autumn of 1865, near the mouth of the Jenissee. An expedition was dispatched to the spot by the Imperial Academy of Sciences of St. Petersburg, in the summer of 1866, and the result of that expedition was the disclosure of some interesting facts in the natural history of a former creation.*

Half-a-ton of this mammoth ivory was sold at the second series of ivory sales in London in May 1876, and realised from £22 to £43 per cwt.

Recent examinations on the Yukon river in Alaska, establish the fact that the fossil-remains of the elephant are even more plentiful on the west than on the east side of the North Pacific. A supply of ivory sufficient to last the world for years has thus been met with.

Ivory is occasionally used now for binding books, but it was often used in older times to preserve valuable missals, &c. There is a carved ivory book-cover of the 11th century in the Museum at Berlin. There is also an elaborate carved one of the same period in the Imperial library at Paris. At Milan is another carved ivory binding to a copy of the Gospels said to be even earlier.†

The scrivelloes or small tusks of the elephant, have been occasionally applied to fanciful ornamental uses by the taste and design of experienced naturalists. Such an instance is afforded in the cut given on the opposite page of ivory tusks mounted for the frame-work of a mirror.

Not much use is made in commerce of the elephant's skin. It has been occasionally tanned as a matter of curiosity. Riding whips of bleached elephant hide were made for the Duke of

* Mr. Lumley's Consular Report on Russia, 1867.

† Cases 171 and 172 contain various applications of ivory, sections of mammoth ivory, sections showing diseased ivory, Chinese carvings, chessmen, draughtsmen, turned balls, fans, &c.; Indian ivory inlaid-work, knife-handles, tablets, brush-backs, combs, hand-mirrors, and other objects of European turnery work, and handles of knives made from the molar teeth of the elephant.

Edinburgh, which had a resemblance to transparent horn. The stout bristly hair on the elephant's tail is made into rings, bracelets, and other female ornaments for the natives of India.

In Bruce's "Travels," it is stated, "The Abyssinians cut the whole of the flesh off the bones of the elephant into thongs, like the reins of a bridle, and hang these like festoons upon the branches of trees till they become perfectly dry, without salt, and

ORNAMENTAL APPLICATION OF SCRIVELLOES OR SMALL TUSKS.

then they lay them up for their provisions in the season of the rains." Sparmann saw the flesh dried in a similar manner by the Hottentots and the Boshjesman race.

CHAPTER XII.

PACHYDERMS, OR THICK-SKINNED ANIMALS, AND THEIR PRODUCTS.

Swine—Great value of the pig—Various breeds—Our foreign imports of pigs, bacon, hams, and pork—Extent of pig-breeding in the United States—Vast export products—Immense slaughter-houses —Pigs in Russia—Pork as food—Trade in lard and lard-oil—Hogskins—Bristles—Our import trade—Russia the chief source of supply—Statistics of Swine in various countries—Peccaries or wild hogs—Tapir—Rhinoceros—Use of its hide and horn—Hippopotamus—Its hide, flesh, and teeth.

SWINE.—The pig is, of all domestic animals, that which has been most modified. The discoloration of the skin, the lessening of the hair on it, the aptitude for rapid growth and fattening, are the results of a system of permanent confinement and rich and abundant food to which the animal has been submitted for many centuries. It is killed at from one to two years, according as there is sufficient food to supply it. After three years the flesh becomes hard, and the animal gains little in size. According to age, race, and degree of fatness, the pig when killed may weigh from 160 to 600 lbs., or more in some improved breeds. It is the animal in which there is the least waste between the dead and living weight, nearly all the carcase being utilised, the blood, the skin, the head, and most of the entrails, which are useless in other animals, serving as food. In highly fattened pigs, the fat constitutes about 27 per cent. of the weight of the animal. The extent and variety of the commercial products of the hog are perhaps unsurpassed by any other domestic animal.

His fat is an article in the form of lard, extensively used in cookery, perfumery, and medicine ; his flesh is prepared for food, either fresh, pickled, as salted pork, dried and smoked bacon; as hams and shoulders usually smoked; the head and collar is made into brawn; the intestines are used as coverings for sausages; the skin or hide is the most general cover for saddles, and is made into shoes and pocket-books; the bladder finds use for excluding air from jars or other vessels, for holding lard, &c.; the bristles are extensively used for tooth-brushes, and for brushes of all shapes and purposes; in short no part of the hog is useless. The very hoofs make jelly or glue, and when to this is added the fact, that his flesh when fresh, is very digestible and nutritious, it forms a catalogue of useful products, not yielded by any other animal.

Exclusive of our home produce, the foreign supply of pigs and their products in 1875 were as follows:—

			Value. £
Live pigs, No. 72,170			255,076
Bacon and Hams	cwts.	2,638,875	6,982,470
Bristles	lbs.	2,558,996	419,203
Lard	cwts.	540,244	1,634,769
Pork	,,	266,663	590,356
			£9,881,874

Besides probably some prepared pig-skins not enumerated.

The enormous increase in the consumption of foreign bacon and hams is worth special notice. The imports have been as follows:—

	Bacon and Hams. cwt.	Pork. cwt.
1840	6,181	29,532
1850	352,461	211,254
1860	326,106	173,325
1870	567,164	257,014
1875	2,638,875	266,663

The varieties of the pig are very numerous, and are increased daily, arising from crossing of breeds, food, climate, care, &c. No animal is so extensively diffused over the globe, or increases so rapidly. Marshal Vauban calculated that the produce of a single sow in ten years, assuming six pigs at a litter, and excluding the males, would increase to 6,434,130 pigs, or as many as any of the chief European States could support. If this calculation were carried on to the twelfth generation, we should find they would people all Europe, and by the sixteenth cover the entire globe.

THE OLD ENGLISH BREED OF HOG.

The Berkshire has long been regarded as one of the superior breeds of England, combining size with a sufficient aptitude to

THE BERKSHIRE.

fatten, and being well fitted for pork and bacon. It has been regarded also as the hardiest of the more improved races.

The extremes of domestic swine are the English prize-pig at the one end, and the pig whose domestic hearth is in the hut of the Finn, all the way from St. Petersburg to Archangel, on the other. This latter is a meagre bony animal only useful for his bristles and his skin.

On most parts of the Continent English pigs have been introduced and become thoroughly appreciated. The different breeds of the numerous English races are now distinguished into black and white, and these again subdivided into large, medium and small.

The wild boars of Barbary, Bengal and Scinde, are much finer animals than those which endure the severity of a northern winter in the forests of Germany or Russia.

Nature made the pig an animal of great activity and spirit. Man, in the due exercise of the power which has been conferred on him of moulding nature to his own convenience, has made him a creature of flitches and hams. This transforming power

SMALL BREED WHITE PIG, SHOWN AT BEDFORD.*

has been exercised rather wantonly, for of all the loaded animals which deform our cattle-shows none so outrage delicacy as the im proved pig. Unless his legs shrink under the weight of his shape-

* From the *Agricultural Gazette*.

less carcase; unless his belly trails on the ground; and unless his eyes are quite closed up by fat, he has no chance of a prize.

North America is the land where breeding swine has been carried out to the greatest extent. The vastness of the capital engaged in pig-raising in the United States, may be estimated from the fact that there were assumed to be, in 1873, 33,630,050 swine, valued at £26,750,000; and the business is fast increasing.

The total product of pork and bacon was:—

	lbs.
	4,484,273,334
Consumed by the population	3,608,023,535
Surplus for sale	876,249,799
Products of the hog exported	503,029,321
Leaving a surplus unutilised of	373,220,478

Among the principal breeds kept are the Suffolk and Chester whites, and the Essex and Berkshire black.

To give an idea of the progression of this trade, I append the exports from the United States for a series of years.

	Hogs. No.	Pork. Barrels.	Bacon and Hams. lbs.	Lard. lbs.
1820-21	7,885	66,647	1,607,506	3,996,561
1830-31	14,690	51,263	1,477,446	6,963,516
1840-41	7,901	133,290	2,794,517	10,597,654
1850-51	1,030	165,206	18,027,302	19,683,082

Shipments of pork from the United States in the last few years:—

	lbs.
1870-1	39,250,750
1871-2	57,169,518
1872-3	64,147,461
1873-4	70,480,384
1874-5	55,152,208

The exports of pork from the port of New York were in

	lbs.
1872	158,194
1873	197,445
1874	178,070

The bacon and hams exported include also a small quantity of beef. It was stated to be in

	lbs.
1872-3	426,986,933
1873-4	373,441,856
1874-5	298,524,240

It is not so very many years ago that the trade in the products of the hog was commenced on a small scale by a few bold speculators in Cincinnati. Selecting the hams and sides of the animal, they made pickled pork; of the rest they took small account. Soon, however, the idea occurred to one more acute than his fellows, that the heads and the feet—nay, even the spine and the vertebræ might be turned to account. Trotters and cheeks had their partizans, and these parts looked up in the market. About this time the makers of sausages caught the inspiration; they found these luxuries saleable; and so many pigs were to be slaughtered, that the butchers were willing to do it for nothing, that is to say, for the perquisite of the entrails and offal alone.

The next step was due to the genius of France. A Frenchman established a brush manufactory, and created a market for the bristles; but his ingenuity was outdone by one of his countrymen, who soon after arrived. This man was determined, it seems, to share the spoil; and, seeing nothing else left, collected the fine hair or wool, washed, dried, and cured it, and stuffed mattresses with it. But he was mistaken in thinking nothing else left. As but little was done with the lard, they invented machines and squeezed oil out of it; the refuse they threw away. Mistaken men again!—This refuse was the substance of stearine candles, and made a fortune to the discoverer of that secret. Lastly came one who could press chemistry to the service of mammon. He saw the blood of countless swine flow through the gutters of the city; it was all that was left of them, but it went to his heart to see it thrown away. He pondered long; and then, collecting the stream into reservoirs, made prussiate of potass from it by the ton. The pig at last was used up.

If we look at the progress of the pig slaughtering trade in Cincinnati, we are struck by the remarkable progress made. In 1833 only 85,000 hogs were killed there; in 1847 the number had increased to 250,000; in the season of 1862, 598,452 were killed, the average weight of each being 203 lbs. and the yield of lard from each 25 lbs.; in the season of 1872 it was 630,301.

The Mississippi Valley is the great seat of the trade in pork-packing which takes in the States of Illinois, Ohio, Missouri, Indiana, Iowa, Kentucky, and a few others. The average number killed during the winter season (which extends from November 1 to March 1) is about 5,500,000.

The chief slaughtering towns are Chicago, Cincinnati, St. Louis, and Indianopolis. If we take the largest central town—Chicago, we find that the hogs received in 1875 were 3,912,110 against 4,259,629 in 1874; those received killed numbered 1,781,876, the average net weight of the hogs was 21,242 lbs. The total yield of lard obtained was 197,038 tierces, being an average of 37·30 lbs. per hog. There were shipped 89,646,605 lbs. of bacon middles; 36,015,801 lbs. of hams, and 31,344,345 lbs. of green and dry salted shoulders. The price of the bacon hogs ranged from 6½ to 7½ dollars each. The total value of the hogs received in Chicago in 1875 was nearly £13,700,000.

There are two seasons termed respectively the winter packing and the summer packing. The following shows the number packed in the winter season for the several States which carry on this business:

States.	Number.
Ohio	886,264
Illinois	1,834,218
Indiana	610,482
Kentucky	333,706
Tennessee	39,860
Missouri	890,679
Iowa	350,087

States.	Number.
Kansas	49,179
Nebraska	20,835
Wisconsin	25,320
Minnesota	331,635
West Virginia	4,000
Michigan	54,989
Pittsburgh, Pa.	20,000
Atlanta, Ga.	4,750
Total	5,456,004

The average gross weight of the pigs was 289½ lbs.; the mean price per 100 lbs. gross $3.78. The aggregate gross weight was 1,579,568,854 lbs. and deducting the usual abatement of 20 per cwt., 1,263,655,084 lbs. To this has to be added the summer packing stated at 118,675,914 lbs.

At Cincinnati in 1866 the average net weight of each pig, without the entrails, was 238 lbs.; 354,079 were brought in alive, and at Chicago 1,410,520; and at St. Louis 191,899.

The slaughtering and packing establishments at Chicago are on an extensive and very complete scale and worked by steam. With incredible rapidity the animal is lifted, killed, washed, scraped, brushed, gutted, cleaned and dismembered. Each part salted or smoked is then put into boxes or barrels, and is ready to be sent into commerce. From 1000 to 2000 pigs can thus be despatched daily in single establishments. These operations can be carried on throughout the year, except at periods of excessive frost.

The building and appurtenances calculated for dispatching two thousand hogs per day, are thus described in an American agricultural periodical:—The hogs, being confined in pens adjacent, are driven, about twenty at a time, up an inclined bridge or passage, opening by a doorway at top into a square room, just large enough to hold them; and as soon as the outside door is closed, a man enters from an inside door, and, with a hammer of about two pounds weight, and three feet length of handle, by a single blow aimed between the eyes, knocks each hog down so that scarce a squeal or grunt is uttered. In the meantime, a second apartment adjoining this is being filled; so the process continues. Next a couple of men seize the stunned ones by the legs and drag them through the inside doorway on to the bleeding platform, where each receives a thrust of a keen blade in the throat, and a torrent of blood runs through the latticed floor. After bleeding for a minute or two, they are slid off this platform directly into the scalding vat, which is about twenty feet long, six wide, and three deep, kept full of water heated by steam, and so arranged that the temperature is easily regulated. The hogs, being slid into one end of this vat, are pushed slowly along by men standing on each side with short poles, turning them over so as to secure uniform scalding, and moving them outward so that each one will reach the opposite end of the vat in about two minutes from the time it entered. About ten hogs are usually passing through the scalding process at one time. At the exit end of the vat is a contrivance for lifting them out of the scalding water, two at a time, unless quite large, by the power of one man operating a lever, which elevates them to the scraping table. This table is about five feet wide and twenty-five feet long, and has eight or nine men arranged on each side, and usually has as many hogs on it at a time, each pair of men performing a separate part of the work of removing the bristles and hair. Thus, the first pair of men remove the bristles only, such as are worth saving for brush-makers, taking only a double handful from the back of each hog, which is deposited in a barrel or box. The hog is then given a single turn onward to the next pair, who, with scrapers, remove the hair from one side, then turn it over to the next pair, who scrape the other side; the next scrape the head and legs; the next shave one side with sharp knives, the next do the same to the other side, and the next the head and legs—and each pair of men have to perform their part of the work in twelve seconds of time, or at the rate of five hogs in a minute, for three or four hours at a time! Arrived at the end of this table with the hair all removed, a pair of men put in the gambril stick, and swing the carcase off on the wheel. This wheel is about ten feet in diameter, and revolves on a perpendicular shaft, reaching from the floor to the ceiling, the height of the wheel being about six feet from the floor. Around its periphery are placed eight large hooks, about four feet apart, on which the hogs are hung to be dressed; and here again we find remarkable despatch secured by the division of labour. As soon as the hog is swung from the table on to one of these hooks, the wheel is given a turn one-eighth of its circuit, which brings the next hook to the table, and carries the hog a distance of four feet, where a couple of men stand ready to dash on it a bucket of clean water, and scrape it down with knives, to remove the loose hair and dirt that may have come from the table. The next move of the wheel carries it four feet further, where another man cuts open the hog almost in a single second of time, and removes the larger intestines, or such as have no fat on them worth saving, and throws them on to a large table behind him, where four or five men are engaged in separating the fat and other parts of value—another move and a man dashes a bucket of clean water inside, and washes off any filth or blood that may be seen. This completes the cleaning or dressing process—and each man at the wheel has to perform his part of the work in twelve seconds of time, as there are only five hogs at once hanging on the wheel, and this number are removed, and as many added, every minute. The number of men employed, besides drivers outside, is fifty; so that each man may be said to kill and dress one hog every ten minutes of working time, or forty in a day. At the last move of the wheel, a stout fellow shoulders the carcase, while another removes the gambril-stick, and backs it off to the other part of the house, where they are hung up for twenty-four hours to cool, on hooks placed in rows on each side of the beams just over a man's head. Here are space and hooks sufficient for two thousand hogs, or a full day's work at killing. The next day, or when cool, they are taken by teams to the packing-house, where the weighing, cutting, storing, and packing, are all accomplished in the same rapid and systematic manner.

The importance of the products of the hog in America are shown by the exports in 1875 which were as follows, from the principal American ports and Montreal.

Pork, barrels	223,803
Bacon and hams, lbs.	184,502,606
Lard, lbs.	166,402,584

There was also made 8,552,583 gallons of lard oil in the States, worth 7,701,900 dollars.

Passing now to another great pig producing country, we find that as regards the number of swine, Russia stands first among the European States, the German Empire, Austria, and France, ranking next in succession. The indigenous races reared in Russia are the short-eared or native, the long pendant-eared, the tridactyle pig, and the breeds of Caucacus and Siberia.

The short-eared race is not a beneficial one. Its head is massive, legs long, and body covered with hair; an adult pig furnishing about one pound of bristles. It fattens but slowly and does not attain its full development till its third year, gives a fair quality of lard, but the flesh is meagre and bad. Its ordinary weight when fattened is from 200 to 360 lbs. The Finnish or lop-eared pig is principally raised in the western part of Russia. It exceeds the common short-eared race in size and will fatten up to 540 and 720 pounds. It is more costly to fatten but yields in return a better quality of meat and a larger amount of lard.

The pigs of the Caucasus have also long pendant ears, and are generally of a black colour. They are very prolific, the sows producing 10 to 18 at a litter. In the Russian villages the pigs are left to roam at perfect liberty, seeking their food in the woods, fields, or streets, and it is only in rigorous winters that they are enclosed in pens and fed. Fattening is only carried on in the vicinity of large manufactories of starch, beet-root sugar, and oil, or distilleries.

Of late years there has been an export of swine carried on from Russia. The average amount in numbers between 1869 and 1872 was 336,205; but in 1873 it rose to 689,000, and in 1874 to 845,000, valued at about 30*s.* each.

The consumption of all the products of the hog is very large in France, the business of the *charcutier* being much more extended than in England, and of these shops there are no less than 700 in Paris. At the annual ham fair held in that city on three days in Lent, the quantity of bacon, hams, and sausages, brought for sale reaches an amount of 500 tons. The number of pigs sold in the Paris markets is about 250,000 annually.

Pork is the meat principally consumed in France and most of the countries of Europe. It requires all the care and judgment of the producers, however, to rear a variety of pig which combines a quantity of good flesh with firmness of fat suited to form the nutritive food of man at the least possible outlay.

Most of the country people eat only pork. Some localities will not touch mutton, and others do not like beef, but there is not a village, a hamlet, or a cottage, where pork is not the ordinary food every day of the year. It is with pork that the daily soup is made, and with its fat or lard that nearly all vegetables are cooked. The peasant prefers to sell his milk and butter, and depend upon the fat of his pig for the wants of his household. There are, in France, two or three million heads of families or households, and it is within the mark to state that nearly every one of these consumes at least, one pig a year, indeed many kill three, four, or five a year; hence instead of there being only five or six million pigs, it is probable that there are double that number. The French race of pigs is found to suit best the special wants of the people, producing a firmer fat, which does not waste in cooking, and hence is better fitted for the requirements of the French kitchen. In the French jury reports of the Exhibition of 1867, M. Reynal, the reporter, asserts that in a pig weighing 160 lbs. the English pig will only yield 50 lb. of meat, while a French pig of the same weight would give double, and better suited to the several purposes of the pork butcher.

The famous Hungarian pork sausages or salami, as big as a man's arm, are largely consumed throughout the Austrian empire. Hungary supplies a great quantity of bacon and lard to the whole of Europe. But the pigs run to fat very much and are long in the leg, and many of them are so useless for food purposes that in some establishments in Pesth half-a-million are annually boiled down for their lard alone.

Lard.—The melted fat of the pig forms a very extensive article of commerce as well as of home consumption in many countries.

The melting of lard preserves it without using salt or any other antiseptic, simply because the pure lard is contained in very minute meshes or cells, and this cellular membrane being the parts that putrify, any process which will destroy or burst them while they are yet sweet, so that the lard may be collected separately, is all that is necessary for its curing or preservation.

The application of heat has been found the simplest and most effective mode of bursting these cells. Melting in a cast-iron pan is the most usual way; but in consequence of the liability to scorch or singe, melting by steam has been lately introduced, but steamed lard is not so certain to keep sweet, owing, no doubt, to the cells being imperfectly burst; the heat of the steam not being intense enough for the purpose. Fire melting can be managed with entire safety under a proper system.

Bladders are a good deal used for holding small quantities of lard. Pigs' bladders are the largest produced and run in more uniform sizes than calves' and bullocks' bladders. It is important that the bladders should be well cleaned by scraping and the use of acids, so that they may be tolerably transparent.*

* A good collection of bladders and their uses is shown in the Waste Products Collection of the Museum.

The leaf-lard and trimming, together with the heads and fat and such other parts as are not useful for packing, are placed in large high tanks and heated by high pressure steam for some hours, until the lard is completely set free. The lard is then drawn off, clarified, and cooled; the leaf-lard is, of course, kept separate from the rest, being much more valuable. The second quality of lard is used for the production of lard oil, while the third quality or that made from such trimmings as have become slightly tainted, is used in the production of the lower grades of oil or for making soap. All hogs that die on the railway cars are also rendered and converted into the lower grade of grease or grease-oil. The refuse bone and meat from the rendering tanks, the internal organs, and the blood, are used in the manufacture of fertilisers.

The exports of lard from the United States have been as follows in the past five years.

	lbs.
1870–71	80,037,297
1871–72	199,651,660
1872–73	230,534,207
1873–74	204,327,536
1874–75	166,859,168

Most of this is shipped from the port of New York.

The exports from that port were in 1872, 173,736,353 lbs.; in 1873, 184,175,568 lbs.; and in 1874, 139,982,979 lbs.

The greater part of this lard is sent to England and the West Indies. The quantity received in the United Kingdom has been yearly increasing. The imports were, in

	cwts.		cwts.
1841	6,226	1871	477,568
1851	120,409	1875	540,244
1861	324,691		

The value of the lard received here in 1875 was £1,634,769.

Much of the American lard is exported to the Havana where it is used instead of butter. The lard oil is said to be used in the

Eastern States to adulterate spermaceti oil and in France to lower the price of olive oil, from 65 to 70 per cent. being often mixed with the olive oil without detection. The impure lard and the fat extracted from diseased animals and from the offal is used in the manufacture of soap.

Lard Oil is now made to a considerable extent, and has many economic uses. The lard in woollen bags is placed between the plates of immense hydraulic presses, and left about 18 hours under a pressure of from 100 to 150 tons. By this means the oil is expressed in a very pure and fluid state.

In Cincinnati alone there are more than forty manufactories of lard oil, which make about 1,500,000 gallons of oil, and 5,200,000 lbs. of stearine for candles. Glycerine is also now largely made. The lard is heated with water at 662° to 720° Fahr., by which the glycerine is separated from the fatty acids, and about 500,000 lbs. are made annually by two or three establishments.

Hogskins, besides their use for saddles, are prepared for leggings, portmanteaus, bags, and other purposes. The best curried fetch from 16*s.* to 30*s.*, the second quality 13*s.* to 15*s.*

Bristles.—After wool and silk one of the most important of the animal fibres we import is bristles. From 1840 to 1854, the average quantity of bristles imported into Great Britain ranged from 1,533,000 lbs. (the lowest), in 1842, to 3,237,000 lbs. (the highest), in 1853. Since that period, the average quantity received has been about 2,500,000 lbs. We have begun even to import bristles from such a distant region as China. The declared value of the bristles annually imported into the United Kingdom in 1874 and 1875 was £420,000. The supply of bristles comes not in bulk from the United States or any part of America, north or south, where they want flexibility—not from Great Britain, where swine by intermixing and crossing, have lost in the value of their clothing what they have gained in the delicacy of their flesh—not from France or southern Europe—nor from many of the

German States where they yield, in the one case, too long a bristle, in the other too short and rigid, and hence have for many years been thrown out of the market—not even from Poland, whose immense woods furnish mast and acorns for herds of swine, but whose bristles, once in demand, either from natural deterioration or careless curing, have now fallen dead upon the market. The trade, indeed in bristles, is both American and English; and these two nations hold its monopoly, but not its supply. It is upon Russia that the whole world is now mainly dependent for this most important household necessary and valuable article of commerce. In the interminable forests of Northern Muscovy, where, for thousands of square miles, the country is covered with pine and larches, oaks and beeches, birch trees and rowans, and the ground littered knee deep with cones, acorns and berries, the swineherd tends his half wild animals, that swarm in almost incredible numbers, and drives them to water and feed. But not all of these are of value for their clothing. The perfect bristle is found only upon a special race, and that race fattened in a certain way. On the frontiers of civilisation, all over the Muscovite territory, are large government tallow factories, where animals, reared too far from the habitations of men to be consumed for human food, are boiled down for the sake of their fat. The swine are fed on the refuse of these tallow factories at certain seasons, and become in prime condition after a few months' feeding. It is from these animals that the bristles of commerce mainly come.

Russia has long been celebrated for the quantity and quality of its bristles. The best are obtained from the common rustic pigs, those in fact which approach nearest to their savage parent, the wild boar. The finest are those taken from the dorsal spine of the animal. The largest quantity of bristles are obtained from the central governments, principally those of the north. Those of Siberia and of Sarapoul in the government of Viatka, are noted for furnishing the best. The average price is from £7 10*s.* to

£8 5s. the poud of 36 lb. The average amount exported reaches 114,440 pouds of the value of about £1,000,000.

France furnishes to commerce a large quantity of bristles, about 2,000,000 lbs. from her native swine (about 850 grammes from each pig); this includes the bristles obtained by plucking and scalding hogs in Midi, Brittany, and La Champagne, the total value being estimated at about 1½ million francs.

French bristles, whether produced in France or only cleansed there, bear the highest reputation in the market. They are white as wool, soft as an infant's hair, firm to the touch and exceedingly elastic. From these are made the brushes of the artist, and the pencils of the skilled painter and decorator. Bristles arrive in casks. They are tied in bundles, and packed with care. Length, elasticity, firmness and colour are the elements that constitute their excellence. The bristle expert selects those of six inches long to make up his class. Above seven will not do. The nine and ten inch bristles lack toughness. Length being settled, elasticity is next looked for. A single brush upon the hand and a glance into the bundle are enough. Colour then comes in. The dark go into one class, the brown to a second, the white to a third. Bristles may be bleached by remaining two or three days in a saturated solution of sulphurous acid in water; most kinds can be bleached by merely moistening them and exposing them to the air, or by moistening them with very dilute sulphuric acid and exposing them to the sun. It is curious that what is considered the meanest employment for the article should have put value upon the choicest selection. The cobbler's bristle rises in value above all others. The price of bristles steadily advances, the range being from 1s. to 7s. per lb.; the demand is beyond the supply.

At Konigsberg, in 1870, the transactions in bristles amounted to 500,000 lb.; half of these were produced in the province, and half came from Russia. About 100 men and women are employed there in assorting the raw material into the following classifications: First and second grey crown, first and second white crown, white and grey shoemakers', and white and grey long.*

Our imports of bristles in 1875 came from the following countries, in the proportions named:—

	lbs.	Value. £
Russia	1,243,966	189,610
Germany	769,873	149,174
Holland	288,127	43,581
Belgium	44,643	8,180
France	99,109	15,239
United States	38,730	4,807
Other countries	74,548	8,612
	2,558,996	419,203

Statistics of SWINE in various countries according to the latest official returns.

Country	Year	Number
EUROPE.		
Russia	1876,	11,694,000
Sweden	1873,	421,795
Norway	1865,	100,000
Denmark	1874,	450,000
German Empire	1873,	7,124,088
Saxony	1867,	325,564
Holland	1873,	360,258
Belgium	1866,	632,301
France	1876,	5,889,000
Portugal	1870,	776,868
Spain	1865,	4,264,817
Italy	1874,	1,574,582
Austria Proper	1871,	2,551,473
Hungary	1870,	3,692,788
Switzerland	1866,	304,428
Greece	1867,	55,776
Great Britain	1876,	2,293,717
Ireland	1875,	1,249,235
Malta and Gozo	1860,	4,500
AFRICA.		
Cape Colony	1875,	110,489
Natal	1875,	11,317
Réunion	1874,	71,490
Mauritius	1875,	30,318
AMERICA AND WEST INDIES.		
United States	1875,	28,062,200
British America	1871,	1,425,014
Uruguay	1874,	100,000
Argentine Republic	1870,	170,000
French Guiana	1874,	5,311
Falkland Islands	1875,	6,000
Jamaica	1870,	9,086
Other British West India Islands	1874,	20,000
Guadaloupe	1874,	12,123
Martinique	1874,	15,352
AUSTRALASIA.		
New South Wales	1875,	219,958
Queensland	1874,	42,884
Victoria	1875,	137,941
South Australia	1875,	78,019
Western Australia	1875,	13,290
Tasmania	1875,	51,468
New Zealand	1874,	123,741

* Case 95 contains a good assortment of the bristles of commerce and of the other hairs used in brush-making, such as badger, camel, goat's hair, etc., while in Case 96, applications of the bristly spines of the porcupine are shown in a writing stand, pen holders, etc.

The wild hogs of Brazil (*Dicotyles labiatus* and *D. torquatus*) live in herds of sometimes as many as sixty individuals. The flesh is one of the best of Brazilian game.

The peccaries of America are, however, much inferior to the wild hogs of Europe and Asia as an article of food, and are said to be only made palatable by the removal, immediately after death, of a singular dorsal gland which is found above the posterior vertebræ.

The hide of the TAPIR (*Tapirus Americanus*), when tanned, makes excellent boot soles, and is highly prized by the Indians for the manufacture of shields. The durability and resistance of this hide is proverbial. The flesh, when roasted, closely resembles beef, and is even compared to veal, especially if it be young. The fatty protuberance on the nape of the neck is alleged to be a delicacy worthy of the table of a modern Lucullus. The feet and cheeks, boiled to a jelly, are also considered delicious.

RHINOCEROS.—Prof. W. H. Flower, F.R.S., lately read a paper before the Zoological Society on some cranial and dental characters of the existing species of rhinoceroses. It gave the result of the examination of fifty-three skulls of rhinoceroses, contained in the Museum of the College of Surgeons and the British Museum, and described the principal characteristics of the five forms under which they could all be arranged, viz. :—1. *Rhinoceros unicornis*, Linn. (including *R. stenocephalus*, Gray); 2. *Rhinoceros sondaicus*, Cuv. (including *R. Floweri* and *R. nasalis* of Gray); 3. *Ceratorhinus Sumatrensis*, Cuv. (including *C. niger*, Gray); 4. *Atelodus bicornis*, Linn. (including *A. keitlon*, A. Smith); 5. *Atelodus simus*, Burchell. It was also shown that the skull of a rhinoceros, lately received at the British Museum from Borneo, was that of a two-horned species not distinguishable from *C. Sumatrensis*.

The African species have two horns, the Indian, with the exception of the Sumatra species, but one horn.

Dr. Barth mentions having often met with the rhinoceros in Central Africa, but only in the eastern part of the country.

Livingstone notices the Kòuabaoba or straight-horned rhinoceros (*R. Oswellii*) which is a variety of the white (*R. simus*, Burchell). Although four species of the rhinoceros are enumerated by Dr. Smith, Livingstone is of opinion that there are only two, and that the other supposed species consists simply of a difference in size, age, and direction of the horns.

The rhinoceros must at one time have been very numerous in the Cape Colony. There are still a great many in the north-eastern part of Great Namaqualand, the northern part of the Kalitari and Bitchouanaland, and the country along the Limpopo. There are said to be four distinct species:—1. The *Rhinoceros Africanus*, the common black rhinoceros with two horns of unequal length, once roaming in the immediate vicinity of Cape Town; 2. the *Rhinoceros Keitloa* or black rhinoceros with two horns of nearly equal length; 3. the *Rhinoceros simus* or common white rhinoceros; and 4, the *Rhinoceros Oswellii*, or long-horned white rhinoceros, the most rare of all.

Of the hide of the rhinoceros (as well as of the hippopotamus) the Dutch Boers make a sort of horse-whip known by the name of jambok. They first of all cut the hide into long strips three inches in breadth which are hung up to dry with a heavy weight appended to them. When thoroughly stretched and dry, these strips are again cut into three divisions, then tapered and rounded with a plane, and a polish given with a piece of grass, which renders them semi-transparent like horn. A jelly from the skin is much esteemed in China and Siam, and a good gelatine is also made with the feet of the rhinoceros.

The illustration on the next page represents a carved rhinoceros horn from Siam, shown at the London International Exhibition of 1862.

There are in the Animal Products Collection of the Bethnal Green Museum, in Case **169**, sections of rhinoceros horn and examples of its commercial applications formed into tazzas, umbrella handles, walking sticks, whips, and other uses.

There is also a good skull with the horn *in situ* of the Indian rhinoceros and three specimens of horns of the rhinoceros of

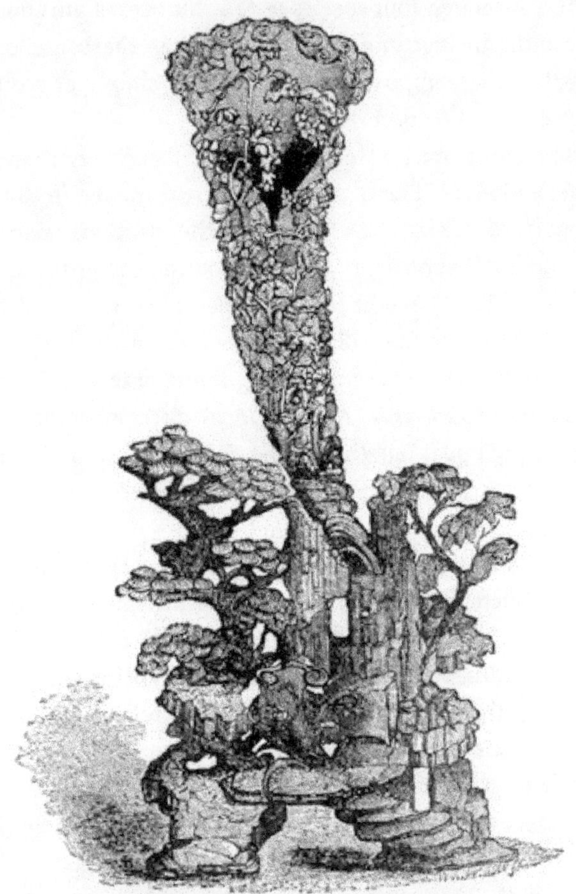

CARVED RHINOCEROS HORN, SIAM.

different sizes. These horns are not attached to the bone, but merely rest upon it. They are composed wholly of horny matter, and this is disposed in longitudinal fibres; so that the horn seems rather, as Prof. Owen observes, to consist of coarse bristles compactly matted together in the form of a more or less elongated sub-compressed cone.

When turned in the lathe and fashioned into drinking-cups, these are held in high repute in Southern Africa by the colonists. The Dutch Boers firmly believe, according to the ancient creed, that if any noxious fluid were poured into a cup of this description, it would instantly foam and boil over the brim. Captain Burton in his "Central Africa" says :—

"The gargetan (karkadan?), or small black rhinoceros with a double horn, is as common as the elephant in the interior. The price of the horn is regulated by its size; a small specimen is to be bought for one jembe or iron hoe. When large the price is doubled. Upon the coast a lot fetches from six to nine dollars per frasilah, which at Zanzibar increases to from eight to twelve dollars. The inner barbarians apply plates of the horn to helcomas and ulcerations, and they cut it into bits, which are bound with twine round the limb. Large horns are imported through Bombay to China and Central Asia, where it is said the people convert them into drinking cups, which sweat if poison be administered in them; thus they act like the Venetian glass of our ancestors and are as highly prized as that eccentric fruit, the *coco de mer*. The Arabs of Muscat and Yemen cut them into sword-hilts, dagger-hafts, tool-handles, and small boxes for tobacco and other articles. They greatly prize and will pay twelve dollars per frasilah, for the spoils of the kobaoba, or long-horned white rhinoceros, which, however, appears no longer to exist in the latitudes westward of Zanzibar island."

A long perfect horn sometimes sells in China as high as £20, but those that come from Africa do not usually fetch above £6 or £7 each. The principal use of these horns is in medicine and for amulets, for only one good cup can be carved from the end of each horn, and consequently the parings and fragments are all preserved.

3½ tons of these horns were imported into London in 1874.

Good sound rhinoceros horns of 28 to 32 inches long will fetch 4*s*. to 5*s*. per lb.; smaller ones, from 13 to 25 inches long, 10*d*. to 1*s*. 6*d*. per lb.

Hippopotamus.—This thick-skinned animal (*Hippopotamus amphibius*) is a native of Africa, and is well known now to most persons by the specimens to be seen in the Zoological Gardens, Regent's Park. It is popularly termed the sea-cow by the Dutch

HIPPOPOTAMUS AMPHIBIUS.

settlers of Southern Africa. The interest attaching to this animal from an economic point of view is rather limited, being chiefly confined to a local use of its skin and flesh in Africa and the limited application of its powerful teeth or tusks as a substitute for true ivory.

Livingstone found the Kafue river (in 16°) full of hippopotami, the young being perched on the necks of their dams. About a thousand of these large animals must have been slaughtered yearly to meet the demand for this ivory.

The flesh of the hippopotamus is delicate and succulent. The layer of fat next the skin makes excellent bacon, technically denominated hippopotamus "speck" at the Cape. Dr. Schweinfurth says that when boiled, hippopotamus fat is very similar to pork lard, though in the warm climate of Central Africa it never attains a consistency firmer than that of oil. Of all animal fats it appears to be the purest, and, at any rate, never becomes rancid, and will keep for many years without requiring any special process of clarifying; it has, however, a slight flavour of train-oil to which it is difficult for a European to become accustomed. The hide is in some parts two inches thick, and not much unlike that of the wild boar. The hide of a full grown hippopotamus is sufficiently heavy to load a camel.* From the hide are made most punishing whips, of which a few have occasionally appeared in the shops of London.

Dr. Schweinfurth states that several hundred Nile whips or kurbatches can be made from the hide of a single hippopotamus, and these are in great demand in Egypt. By a proper application of oil, heat, and friction, they may be made as flexible as gutta percha. The fresh skin is easily cut crosswise into long quadrilateral strips, and when half dry the edges are trimmed with a knife, and the strips are hammered into the round whips, as though they were iron beaten on an anvil. The length of these much dreaded "knouts" of the South is represented by half the circumference of the body of the hippopotamus, the stump end of the whip, which is about as thick as one's finger, corresponding to the skin on the back, whilst the point is the skin of the belly.

Hippopotamus teeth were in former years largely in demand by dentists to make artificial teeth, and as much as 30*s.* per pound was paid for fine tusks; the dentine is very hard and dense, and contains less organic matter than the ivory of the elephant or the tusk of the walrus. Mineral teeth, however, have almost exclusively replaced those made of this substance. Some seven or eight tons of hippopotamus teeth used to be imported yearly, but they fetch a very low price now—1*s.* 6*d.* to 2*s.* 6*d.* per lb. When great delicacy is required, this dentine, however, is found more suitable for carvings than the ivory of the elephant, and is much used in France for this purpose, especially in the manufacture of fine

* There are, in the Animal Collection of the Museum, samples of dried rhinoceros hide and elephant's skin, from Siam, in Case **140**; and strong shooting shoes made of thick boar's hide. And on the south wall a wild boar hide for a shield, which took two years' tanning. Under the stairs, there are part of a tanned rhinoceros hide and a hippopotamus hide.

brooches. Pausanius (viii. 46) mentions the statue of Dindymene, whose face was carved out of these teeth instead of elephant ivory.

The tusks of the hippopotamus, like the incisors of the rodents, are maintained of a certain definite shape by being opposed to one another, but the tusks of the walrus and of the elephant, as it would seem through the absence of opposing teeth, acquire proportionately greater length, exhibiting, indeed, the largest size to which teeth attain in the animal kingdom.

There is a fine skull of the hippopotamus, and several of the canine tusks and incisors shown in the Museum Collection, on the west wall.

CHAPTER XIII.

AQUATIC MAMMALS.

Sirenia—Dugongs—Whales and the whale fishery—Products of the fishery—Changes in the trade—Its great decline—Whalebone—Spermacceti—Ambergris—Whale-oil—The Narwhal or sea-unicorn—The common Dolphin—Porpoise—White Whale—Porpoise-oil—Porpoise-leather.

HERBIVOROUS MARINE MAMMALS.—The eastern representative of the family of *Sirenia* is the DUGONG (*Halicore Indicus*, Desm.; *Trichechus Dugong*, Gmel.; *H. Australis*, Owen and Gray; *Halicore Dugong*, F. Cuvier; *Dugungus Indicus*, Less and Hamilton). It is distributed over the Indian Ocean, common in shoals, on the coasts of Ceylon, Aripore and the Bay of Calpentyn to Adam's Bridge. It is also found along the eastern coast of Australia from the Tweed to Cape York. The dugong is the principal of the herbivorous cetacea, subsisting entirely on a species of sea-grass which grows in those shallow waters, which are its regular feeding grounds.

Professor Owen remarks, "the whole of the internal structure in the herbivorous cetacea differs as widely from that of the carnivorous cetacea as do their habits; the amount of variation is as great as well could be in animals of the same class, existing in the same great deep. The junction of the dugongs and manatees with the true whales cannot, therefore, be admitted in a distribution of animals according to their organization. With much superficial resemblance, they have little real or organic resemblance to the walrus, which exhibits an extreme modification of the amphibious carnivorous type. I conclude, therefore, that the dugong and its congeners must either form a group apart, or be joined, as in the classification of M. de Blainville, with the pachyderms, with which the herbivorous cetacea have most affinity."

The upper lip is very large and thick, and both are covered with a hard horny substance, which aids the animal in tearing up the submarine weeds which are its food. The eyes are small, like most others of the whale tribe, and the ears very minute. They usually frequent the ocean inlets, where their food most abounds, browsing on them like an ox eating up the grass. In size these animals seldom exceed 14 feet in length. The fatty oil is obtained from the adipose matter of the cellular substance under the skin. It is free from any odour, and when well refined, is clear and limpid. At low temperatures it nearly becomes thick, and loses its fluidity.

The dugong is an inhabitant of the Indian Ocean, and is found principally amongst the islands of the Archipelago, and the north-eastern coast of Australia. As far as we are aware, they are not met with south of Moreton Bay, and even there they are not by any means as numerous as in the bays and shallows farther north. In the waters of Wide, Hervey, and Rodds' Bays they are far more numerous. But their chief habitat is the tropics. There they may be found at all seasons of the year in almost incredible numbers, coming in and going out with the tide, just like huge mobs of cattle at a mustering on a large cattle-station. Dugongs being obliged to come to the surface to breathe, are rarely found in deep water, but usually in from two to four fathoms, where grow the grass and weed which are their sole food.

In length they vary from six to fifteen feet, the average probably being about nine feet. The weight of an ordinary specimen will be from four to six hundredweight, although we have heard of individuals weighing something like a ton, and producing no less than twenty gallons of purified oil. Directly behind their peculiar flippers are situated the mammæ, or breasts, which are not of very large proportions. The young are born singly, and are known as "calves." When killed and opened, a dugong is about the size and appearance of a bullock, except that the skin is thicker, the

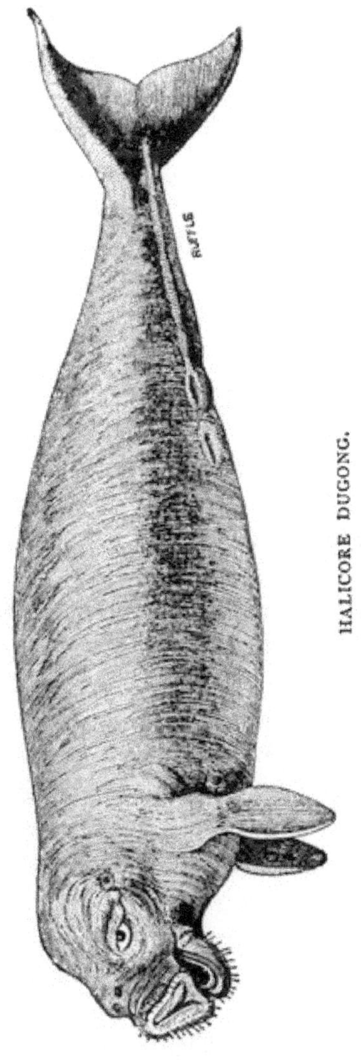

HALICORE DUGONG.

fat more like that of the pig, and that the tail part assumes the shape of that of a fish.

The skeleton of one forwarded to Europe from Australia some years ago measured eleven feet in length.

The flesh is excellent, tasting more like beef than fish. Gumilla states that the flesh of those of the Orinoco, when roasted, has the flavour of pork and the taste of veal, and when salted makes excellent sea-store. The dugong is considered by the Malays as a royal fish, and the king is entitled to all that are taken. The flesh is highly prized, and considered by them as superior to that of the buffalo or cow.

The dugong, or sea-hog, locally called "Mooda Hoora," is sometimes met with in shoals on the western coast of Ceylon. When captured, the flesh of this cetacean sells there at half-a-crown for a small piece of less than a pound in weight. It is esteemed a great delicacy by the Mahometans, who naturally seek a compensation in this dish for the prohibition under which they suffer respecting the porcine type terrestrial.

The flesh of the dugong is much relished by the blacks. Dr. Kelaart, of Ceylon, says it is far from being disagreeable, and is not unlike pork.

They are caught about the Straits Settlements, eight or nine feet long. Leguat, an old author, writing nearly 150 years ago, says they were met with in shoals of 300 or 400 around the Isle of France or Mauritius, and that they were sometimes twenty feet long.

Dugong bacon has become an article of sale at Maryborough, Queensland, and is described as being, when properly cured, a relishing article of diet, more especially for the breakfast table. It is said to have the flavour of good bacon with just an agreeable "bloater" twang added. Mr. E. Wilson states that he unconsciously ate it in salted rashers for breakfast without knowing the difference.

The "Moreton Bay Press" (Queensland) states that the bone of the dugong, or "yungan" (as the aborigines call it), in fineness and hardness of grain, specific gravity, and appearance, approaches nearly to the nature of ivory. It could be used as a substitute for that material for most of its purposes, and might be procured in considerable quantities for export.

The oil from the dugong has obtained a high reputation as a substitute in medicine for cod-liver oil.

Dr. Hobbs, the health officer of Moreton Bay, tells us (what the later experience of many others also confirms) that "so sweet and palatable is the oil procured from the dugong, that in its pure state it may be taken without disagreeing with the most sensitive stomach, and also used in a variety of ways in the process of cooking; so that this potent restorative remedy may be taken as food, and many ounces consumed almost imperceptibly every day, thus furnishing the system with the requisite amount of carbon for its daily oxidation."

The Queensland Dugong fishery, situated in Hervey's Bay, near Maryborough, is owned by Mr. Y. L. Ching, a gentleman who has gone into the business in an extended and systematic manner. The dugong is taken either with nets or by harpooning. Three or more boats are sent out daily, being manned by aboriginal crews (all of whom are picked men), whose sighting capabilities are very superior, far surpassing the whites at this sort of work. In this manner sometimes as many as six dugongs a day are taken. When one has been harpooned, it is immediately brought home and cut up. The oil is then extracted, and subjected to a variety of complicated processes, finally undergoing a chemical clarification, which renders it fit for the European and American markets, where, owing to its medicinal properties, there is a large demand for it. A large dugong weighs from 10 to 12 cwt., and yields from 6 to 14 gallons of oil. Sometimes, indeed, as much as 18 gallons are obtained from a good fish.

Dr. Barth states that a mammal like the *Manatus Senegalensis*, called Ayu by the natives, is met with in the rivers of Central Africa. The South African rivers also have these mammals. The

Ayu is not less frequent in the Isa, near Timbuktu, than it is in the Benuwé.

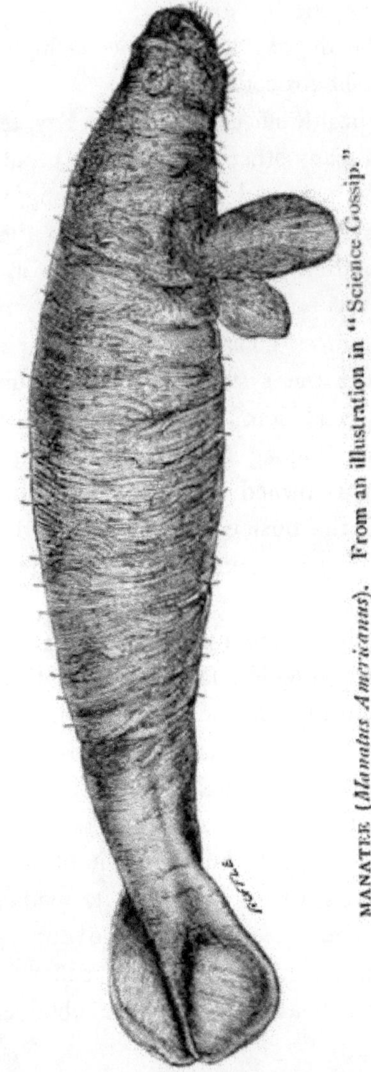

MANATEE (*Manatus Americanus*). From an illustration in "Science Gossip."

There are said to be two Atlantic species—the *Manatus Americanus*, and *M. Senegalensis*. The former frequents the mouths of rivers and quiet secluded bays and inlets among the islands of the West Indies and the coasts of Guinea and Brazil. It is said sometimes to attain nearly 15 ft. in length, and differs from the dugong in having no canines or incisors. It is particularly abundant in the lakes of the Amazon, where it is known as the "peixe boi." Wallace, in his Travels up that river, describes their capture :—" Beneath the skin," he says, " is a layer of fat of a greater or less thickness, generally about an inch, which is boiled down to make oil, used for lighting and cooking. Each yields from five to twenty-five gallons of oil." Edwards, in his " Voyage up the River Amazon," also speaks of them, and says that they are not unfrequently taken eight feet in length. It is said to be a distinct species from the Manatus of the Gulf of Mexico.

The skin of these animals is smooth, of a blackish-blue colour, much stouter than that of the ox, being nearly an inch thick. A Queensland paper recently stated that 49 dugong-skins were shipped by Mr. Smith, the naturalist, from Wide Bay.

THE WEST INDIAN MANATEE (*Manatus Americanus*).*

" Their skins," it observes, " will probably find their way back to many of our mills and factories, in the shape of machine-belting, &c. Owing to the thickness and durability of the hide of the dugong there is an increasing demand for it, which is likely to lead to a considerable trade in this new commodity."

* From Wilson's "Zoology."

They make thongs of it about the Orinoco River, South America, to tie their cattle, and for horsewhips. Canes and sticks of Lamantin (*Manatus Americanus*) are made by moistening the dry layers of the skin, which are taken off when this mammal is caught, and it is very abundant in the seas around Cuba. These are beaten and passed over the fire to round and polish them. They are only made to order, but might become an important industry if the skin of the manatus entered more into commerce, as it is capable of many other applications, such as riding-whips, surgical instruments, &c. The price of the walking-sticks in Trinidad de Cuba is from £2 to £3 10*s.* according to the care with which they are prepared.

WHALES.—These cetaceous animals, of which there are several species, are of commercial importance on account of the oil contained in their fat or blubber, and in the head of some, and of the whalebone or horny laminæ in their upper jaw. Large fleets used to be fitted out by different countries for their capture, but the outfit of vessels has greatly declined in number and tonnage of late years.

One species of whale is met with about the North Pole, two in the Atlantic, and two others in the Pacific. On the coasts of South America another species is found, known as the Southern whale. It passes the summer about the Cape of Good Hope, and in winter is met with about the South American coast. There would seem to be also two species in the Pacific. One in the north is met with in winter near America, in summer close to Asia, on the coasts of Japan and China. The southern species passes the spring near America, and is found in winter in the Oceanic seas. The Greenland whale, the object of the principal fishery, never leaves the icy seas. The waters with which the earth is begirt are still full of the leviathans of the deep. They change indeed their habitation from time to time.

The whales caught about the Eastern Archipelago are of the cachalot or sperm kind. Those to the north of Australia and the west coast are of various kinds, the cachalot, right whale, and humpback.

The total number of whales killed in the northern fisheries in the forty-three years ending 1858 was about 24,000, besides twenty or thirty captured yearly in the shore fisheries of the colonies. Hull formerly took the lead in whaling, and during the period between 1772 and 1859 that port benefited to the extent of nearly 7¼ millions sterling by the whaling trade, upwards of

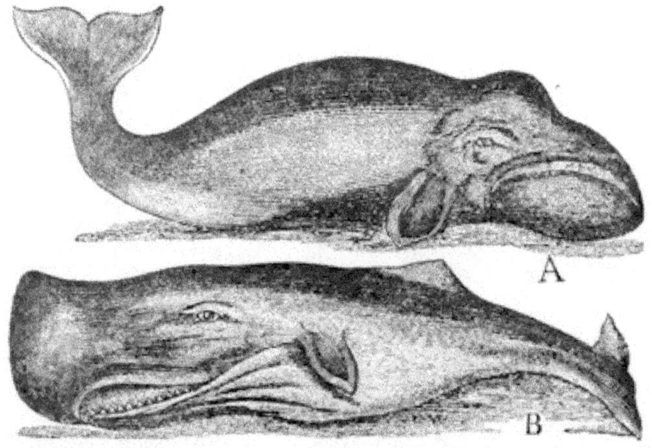

A, RIGHT OR GREENLAND WHALE (*Balæna mysticetus*).
B, CACHALOT OR SPERM WHALE (*Physeter macrocephalus*).

2,000 ships having been sent out at different times. Peterhead has latterly taken the first place in the outfit for the Northern whale fishery. The produce of the fishery brought into that port in the ten years ending 1858 was 1,108 whales captured, yielding 23,636 tuns of oil and 14,455 cwts. of whalebone, of the aggregate value of £1,239,000. The whale trade has been gradually shifting from the ports in this country, which formerly enjoyed the greatest share of the business.

Previous to the year 1790 London was the principal port from whence the vessels sailed, inasmuch as four times as many were sent out from the Greenland Docks at Deptford (now the Commercial Docks) as from any other place. In 1820 55 vessels measuring 17,603 tons were dispatched, which brought in 7,946 tuns of oil, and 19½ tons of bone; in 1830 90 ships were sent out. Liverpool, which also used to fit out many ships for the whale fisheries, has long given up the pursuit and turned her shipping to other channels.

The following is the result of the British whale fishing from Dundee since 1865:—

	No. of Ships.	Oil—Tuns.	Bone—Tons.
1865	7	630	30
1866	11	340	18
1867	11	20	—
1868	13	970	50
1869	10	140	7½
1870	6	760	40½
1871	8	1,165	61½
1872	10	1,010	54
1873	10	1,352	69
1874	10	1,290	66½

The price of whale oil at the commencement of 1877 was £30 to £33 10*s.* per tun. The imports in 1876 were only 45 tuns.

The price of Arctic and Davis' Strait whalebone has run up to the extraordinary price of £1,200 to £1,300 a ton, or about treble the average prices of the past ten years. The consumption of whalebone in this country was, in

	Tons.		Tons.
1872	111	1875	72
1873	104	1876	55
1874	107		

The records of the American whale fishery prove how petroleum and vegetable oils have been driving animal oils from the market. In 1846, 730 ships measuring 233,189 tons, were employed in the American whale fishery. In 1874, only 163 ships measuring 37,733 tons were so occupied; being little more than half the number that were fitted out in 1846, for the North Pacific fishery alone. To show the great fluctuations, I give the imports of the three articles the produce of the American whale fishery at decennial periods.

	Spermaceti. Barrels.		Whale Oil. Barrels.	Whalebone. lbs.
1846	95,217	......	207,493	2,276,930
1856	80,941	......	197,941	2,592,700
1866	36,663	......	74,302	920,375
1874	32,203	.. ...	37,782	345,560

The largest take was in 1853, 103,077 barrels of spermaceti, 260,114 barrels of oil and 5,652,300 lbs. of whalebone.

IN PURSUIT OF WHALES.

Products and exports of the American whale fishery:—

		Spermaceti.			Whale Oil.		
		Barrels.		Exported.	Barrels.		Exported.
1868	......	47,174	......	18,619	65,575	......	9,885
1869	......	47,906	......	18,645	85,011	......	3,842
1870	......	55,183	......	22,773	72,691	......	9,872
1871	......	41,534	......	22,156	75,152	......	18,141
1872	......	45,201	......	24,344	31,075	......	1,528
1873	......	42,053	......	16,238	40,014	......	2,153
1874	......	32,203	......	18,675	37,782	......	3,300

France has quite given up the whale fishery. Ten years ago

two or three vessels used to bring in 200 or 300 tuns of oil and 8 or 10 tons of whalebone, but there have been no vessels employed since 1868.

A coast fishery, for the capture of the whales that approach the shore, is still prosecuted in some few of our Colonies. There is an export of whale-oil from New Zealand.

The whale fishery from Tasmania employed 16 vessels in 1874, instead of 18, as in the previous year. These brought into port 342 tuns of sperm oil valued at £30,780; whereas, in 1873, 558 tuns of sperm oil and 12½ tuns of black oil, valued at £44,000, were returned as the produce. The fishery was carried on in 1875 by 13 vessels. Seventeen vessels obtained, in 1869, 13 tuns of black oil and 643 tuns of sperm, with 8 cwt. of whalebone, the whole valued at £49,000. In 1875 the exports from the colony were :—

			Value. £
Black oil	25 tuns	......	965
Sperm oil...............	314½ ,,	......	29,970
Whalebone	2,000 lbs.	......	38
			£30,773

The largest amount shipped in the past ten years was £52,546 in 1868.

Dr. Scoresby says the flesh of the young whale is of a red colour, and when broiled and seasoned with pepper and salt, eats like coarse beef. The milk of the whale resembles that of the quadrupeds in appearance, and is said to be rich and well flavoured.

Whale-oil.—Blubber is the cellular membrane in which the oil of the whale is included. Its thickness varies from 8 to 20 inches, and frequently as much as 100 tuns of oil is obtained from a full grown whale. The oil is drained out of the blubber by placing the latter cut up in racks, through which the oil drips down into casks. It is then heated up to 225° to deprive it of its rancid smell, and also to make the grosser parts settle. The oil is then pumped over with water, left to cool, and finally stored in casks.

Under the general name of black fish oil of the whalers, several kinds are included. Thus the pilot whale (*Globiocephalus Svineval*) is known as the black whale to seamen, and among American sailors *G. intermedias* is so termed. The common black whale of the Australian Seas is *Balæna Australis*, and the name is sometimes applied to *Physeter microps* and *P. Tursio*. The killer or black fish of the South Sea whalers is the *Globiocephalus macrorhynchus*.

A black fish of average size will produce from 30 to 35 gallons of dark oil, with an unpleasant odour.

Whalebone.—These animals have no teeth, but in their place they have substitutes in the form of baleen plates ending in a fringe of bristles. These two extensive rows of whalebone or "fins," as they are denominated in trade, are suspended from the sides of the crown bone, and hang down on each side of the tongue.

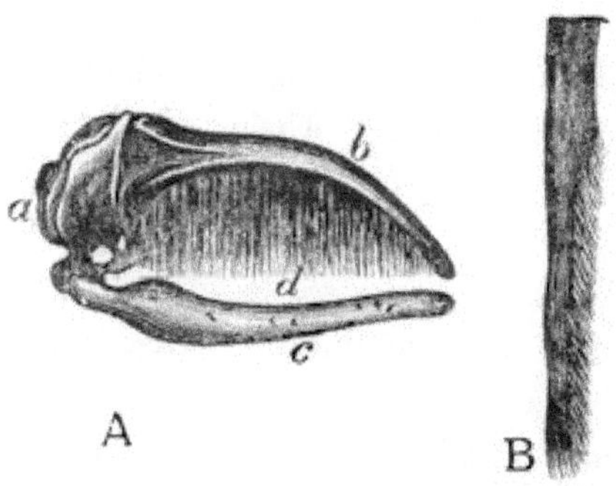

A, SKULL OF WHALEBONE WHALE (*Balæna mysticetus*). *a*, CRANIAL PORTION OF SKULL; *b*, UPPER JAW; *c*, LOWER JAW; *d*, BALEEN.
B, SINGLE PLATE OF BALEEN, SHOWING THE FIBROUS EDGE BY WHICH THE PLATES ARE UNITED TOGETHER.

Each series or "side of bone," as the whale fishers term it, consists of upwards of 300 plates or "blades," as they are commercially called; on account of the arched form of the roof of the mouth, the longest are near the middle, from whence they gradually diminish away at each extremity; fifteen feet is the greatest length of the whalebone, but ten or eleven feet is the average size, and thirteen feet is a magnitude seldom met with. If the largest blade in the series weighs 7 lbs., the whole produce will be about a ton. In young whales or suckers, the whalebone is only a few inches long; but when the length reaches six feet or upwards, the whale is said to be of "size."

Three principal kinds of whalebone are recognised in commerce.

1. The Greenland, from the Davis' Strait fishery and various parts of the North Sea.

2. The South Sea or black fish whale-fin, brought by the South Sea whalers; and

3. The North West coast or American whale-fin, obtained principally in the Pacific or Behring Straits fishery by the United States whalers. These three kinds are very different in shape, and are thus described by the late Dr. J. E. Gray.

"The outer edge of the Greenland is curved considerably, in that of the North-West Coast it is much more straight, and in that of the South Sea nearly quite straight. The fibres on the edge in the Greenland and margined whales are very fine, flexible, and long, forming only a thin series; in the South Sea they are rather coarser; but in the North-West Coast much thicker and coarser, quite bristly, and much more so towards the apex; and they are more erect, and form a thicker series, approaching in that character to the baleen of the 'finners.'

"The Greenland fin has the hair on its edge generally stripped off, and is clean and bright when it is brought to England; but this may be from the care the North Sea whalers take in collecting and cleaning it; the blades are brought home in bundles of about 1 cwt. each. On the other hand, the North-West Coast fin and the South Sea fin have the hair left on the edges: they are brought home in bulk, and are always covered with an ashy-white, soft, laminar coat, looking like the rolled external layers of the enamel. This coat has to be scraped off with large knives before it is used or prepared, and the surface after the scraping is not so polished and resplendent as that of the Greenland 'fins.'"

The baleen of the right whale is very long, varying from nine to twelve feet, linear, tapers very gradually and is of nearly the same moderate thickness from end to end.

The following table shows the receipts, in the United States, of whalebone from the Arctic, Greenland, and Southern Fisheries, and the quantities exported to Europe.

	Receipts. lbs.	Export. lbs.	Consumption. lbs.
1861	1,112,000	1,039,000	105,000
1862	691,000	667,000	97,000
1863	562,000	443,000	60,000
1864	770,000	590,000	152,000
1865	651,000	373,000	202,000
1866	891,000	754,000	195,000
1867	949,000	737,500	132,500
1868	900,850	707,882	256,000
1869	603,603	311,600	167,400
1870	708,365	348,000	210,700
1871	600,655	387,199	213,456
1872	193,793	177,932	115,861
1873	206,396	120,545	85,751
1874	345,560	164,533	181,027

Whalebone is closely analogous, in its chemical and physical properties and mode of growth, to hair and bristles, but is developed in compact plates which resolve into stiff bristles at their free margin, from the thick gum at the circumference and palatal surface of the upper jaw of the animals of the whale tribe. The most valuable kind of whalebone is obtained from the great Arctic whale (*Balæna mysticetus*) in which the plates or blades are arranged in several rows, the outermost consisting of the longest plates, attaining, in a full-grown whale the length of from 8 to 12 feet. The Antarctic whale (*B. Australis*) affords the second best kind, and the great finner whale (*Balænoptera Boops*) the shortest and coarsest plates, which are only fit for splitting into the false bristle for brushes, &c. Prepared whalebone has been applied to covering whip-handles, walking-sticks, and telescopes; and, in the form of shavings for plaiting like straw, in the construction of light hats and bonnets. It has lately been used for the higher qualities of artificial flowers, probably for stamens, &c.

Whalebone has nothing in it of the nature of bone, but is an albuminous tissue, resembling horn and hair, and, as it were, forming the transition between these two substances. This material answers the purpose of retaining the small mollusca and crustacea contained in the water which enters the mouth of the whale, by the fringe-like bristles which project from the central part of the plates. The principal supply is obtained from the Greenland and southern whales (*Balæna mysticetus* and *Australis*). From the roof of the mouth hang down on each side of the tongue about 300 plates of whalebone, or *baleen*, as it is properly called; these are at right angles to the jaw-bone, and the largest of them parallel to each other. On the southern wall of the Museum there are displayed two pairs of blades of the hump-backed whale (*Megaptera longimana* and *M. Americana*) one from Bahia and one from Greenland; one pair of blades of medium size from the South Sea whale (*Balæna Australis*); and one large extra-sized blade, measuring about 12 feet from Davis Strait fishery—this blade has been cleaned and the hair-fringe removed;—one pair of blades from *Balæna Japonica;* and one medium-sized blade of Polar whalebone, cut into sections to show the mode of use, the largest plates being the most external. The form of the baleen-covered roof of the whale's mouth is a transverse arch, in which rests the convex upper-surface of the tongue when the mouth is closed. Each plate, on its inner and oblique margin, sends off a fringe of rather stiff but flexible hairs, which project into the mouth.

The preparation of whalebone consists in boiling it in water for several hours, which renders it soft when hot; this process also renders it harder when it is cold, and of a darker colour; the jet black colour is the result of a dyeing process. Whalebone used to be largely employed for the stretchers of umbrellas, for canes, whips, and as a substitute for bristles in common brushes. It is impossible to use it with advantage on very extensive surfaces, as it cannot be soldered or joined together like tortoise-shell. When softened by heat, it can be bent and moulded, like horn, into various shapes, which it retains, if cooled under the pressure. The surface is polished with ground pumice stone, felt, and water, and finished with dry-sifted quick-lime.

In Case **96** some manufacturing applications of whalebone are shown in plaited work and thin shavings for whips, rosettes, &c., dyed different colours.

Spermaceti is a kind of waxy body which separates in cold weather from the oil obtained from the head matter of the cachalot or sperm-whale (*Physeter macrocephalus*). It is used a good deal for making candles, with about 3 per cent of beeswax added to prevent crystallisation. This manufacture has diminished to a great degree through the invention of composite and other hard candles.

The head of a large sperm-whale will weigh about thirty-five tons, and forty-five barrels of pure spermaceti have been taken from the *case*, which is a mere vein in his head compared with the remaining part, which consists of four-fifths of the head, and is called "white horse"—a sinewy gristle, which is impenetrable to a sharp axe.

In the right side of the nose and upper surface of the head of the cachalot whale, is a triangular-shaped cavity, called by the whalers "the case." Into this the whalers make an opening, and take out the liquid contents (oil and spermaceti) by a bucket. The spermaceti from the case is carefully boiled alone, and placed in separate casks, when it is called "head-matter."

This substance is of a yellow colour, and its consistence varies with the temperature. In purifying it, to separate the oil as much as possible from the solid matter, it is submitted to compression in hair bags placed in an hydraulic press. It is then melted in water, and the impurities skimmed off. Subsequently, it is re-melted in a weak solution of potash. It is then fused by the agency of steam, ladled into tin pans, and allowed slowly to concrete into large, white, translucent, crystalline masses.

Our imports of spermaceti or head-matter have been as follows in the last ten years :

	Tuns.	Value.		Tuns.	Value.
1866	2,121	£263,118	1871	5,388	£451,028
1867	3,226	373,367	1872	3,715	333,534
1868	1,945	185,960	1873	2,817	252,434
1869	4,107	387,171	1874	3,155	296,630
1870	4,069	341,340	1875	4,469	427,884

Ambergris is a concretion from the intestines of the spermaceti-whale, and is a product of disease, as it is not met with in the healthy animal. It is sometimes found floating on the sea, and is occasionally extracted from the rectum of whales in the South Sea fishery. It has a grey white colour, often with a black streak; has a strong, but rather agreeable smell, a fatty taste, is lighter than water, melts at 140° Fahr., dissolves readily in absolute alcohol, in ether, and in both fat and volatile oils. It contains 85 per cent. of the fragrant substance called *ambreine.* This is extracted from ambergris by digestion with alcohol of 0·827, filtering the solution, and leaving it to spontaneous evaporation. It is thus obtained in the form of delicate white tufts ; which are converted into ambreic acid by the action of nitric acid. Ambergris sells at from 21*s.* to 42*s.* the ounce.

Strangely enough, this substance is brought to Mogador in considerable quantities by the Timbuctoo caravans from the interior of Africa. It probably finds its way there from the West coast. At Mogador it sells for about £20 per pound. Most of the well-off Moors have ambergris in their houses, and they use it in green tea as a flavouring, and one of the greatest compliments paid to a guest is to present him with a cup of this curious mixture.

The cachalot or sperm-whale is not unfrequently cast up on the southern coasts of Morocco. One was thus stranded at Casa Blanca in each of the three years ended 1870, and considerable quantities of ambergris are in this way obtained. The exports from Morocco were in:

	lbs.	Value.
1868	27	£600
1869	65	1,300
1870	100	2,000
1873	18	360

Ambergris is occasionally found on the shores of some of the Bahamas Islands; two pieces brought lately (1873) at Nassau over £140 each.

In 1869 ambergris to the value of more than, £7,000 was imported into France from some of the French settlements of Madagascar, Mayotte, &c.

THE NARWHAL OR SEA UNICORN (*Monodon monoceros*).—This is a singular and very interesting animal, formed in many respects like the whale, but much more graceful. The colour is grey above, and pure white beneath; the whole spotted or mottled with a blackish hue. From the head projects a long straight horn of solid ivory, in the same line as the body. The structure and origin of this horn (which has given much celebrity to this handsome creature) are very peculiar. It is, in fact, the tooth, and the only one it possesses in general: the fellow tooth, however, exists within the bone of the jaw, but undeveloped. It is usually the left tooth that projects. Its principal use appears to be to pierce and kill its prey, such as large skates and other fishes.

This ivory sword is seldom used for manufacturing purposes. In the year ending March, 1875, the Greenland Company collected 457 lbs. weight of Narwhal horns which were valued at £175.

The flesh is much esteemed by the Greenlanders and Esquimaux, who eat it dried and smoked, and employ certain parts of the intestines for making strong cords; the tusks are used by them for pointing the extremity of their arrows and harpoons.

The horn of the Narwhal has been sometimes obtained ten feet long; it is spirally striated throughout its whole length, and tapers to a point. It was once in high repute for some supposed

NARWHAL AND POLAR BEAR.

medicinal properties. The narwhal is occasionally, though not very often, found with two of these horns or tusks, sometimes of equal, at others of unequal length. About 100 tons of these are occasionally imported in a year to Denmark from the Arctic seas, under the misnomer of "sword-fish horns." There is said to be a magnificent throne in Denmark made of this ivory, which is preserved with great care in the castle of Rosenburg.

Among the collection of ivory on the east wall, there are four tusks of the narwhal, one six feet eight inches long, and another divided to show the section of the interior ivory. There is also a walking-stick carved from it, a turned walking-stick of the bone of a whale, and a native weapon of similar substance from North West America. In Case **140** is a sample of tanned whale skin.

In Cases **171** and **172** are sections of teeth of the sperm-whale, and some carved.

The COMMON DOLPHIN (*Delphinus delphis*) has at present little or no commercial importance, but in former years a dolphin was thought a fit and worthy present to be made to a Duke of Norfolk, who again distributed it among his friends; it was roasted and eaten with porpoise sauce.

The COMMON OR BLACK PORPOISE (*Phocæna vulgaris*) is the smallest, and the grampus (*Phocæna orca*) the largest of the animals of the Delphinidæ family. The porpoise is commercially the most useful of them.

The length of the common porpoise is from four to eight feet, and when in the water they present a considerable resemblance to large black pigs, whence they are frequently called sea-hogs and hog-fish. The name of porpoise is said to be derived from the Italian *Porcopesce*, or hog-fish; the German Meer-schwein has the same meaning, and the French Marsouin is evidently derived from some old Teutonic form of the same word.

In 1799, when animal oil was more valuable than it is now, the Society of Arts offered a premium of £30 for the greatest number of porpoises beyond thirty which might be taken in the year off the British coasts, and an additional premium of £13 for not less than fifteen tons of oil extracted from porpoises.

The WHITE WHALE.—The extraction of the oil of the white porpoise (*Phocæna leuca*), and of the black porpoise (*P. communis*), constitutes an important industry in the district of Quebec on the St. Lawrence River.

One vessel brought home to Peterhead, in 1868, the oil and skins obtained from 250 white whales.

This oil is inodorous, and gives a brilliant light; it is said to be superior to any other for lighthouses because it does not congeal even in the most intense cold, and its ductility renders it invaluable for greasing leather and oiling machinery, which it preserves from injury by friction.

Porpoises are caught in large numbers in the Little Belt, Denmark, 1100 to 2000 being frequently secured.

The porpoise was occasionally served up at the tables of the old English nobility as a sumptuous article of food, and eaten with a sauce composed of sugar, vinegar, and crumbs of fine bread. It was also often sold salted.

In some parts of North America the skin of the porpoise like that of the beluga is tanned and dressed with considerable care. At first it is nearly an inch thick, but it is shaved much thinner, till it becomes somewhat transparent, and is then made into articles of wearing apparel; it also supplies excellent coverings for carriages.

The fishing for the porpoise is carried on off the coast of Trebizond; it is taken in nets, and is also shot. It yields a large quantity of oil per annum; a portion is consumed by the lower classes for light, and the rest finds a market at Constantinople. The natives at Lazistan are very expert in this business; it forms the occupation of 1,500 to 2,000 individuals.

Porpoise Leather.—Under the name of porpoise leather, the tanned skin of the white whale (*Beluga catodon*, Gray, *Phocæna leucas*), has become an important article of trade. Originally introduced by Mr. W. Tetu of Quebec, in 1851, they met with but little favour from tanners and leather dealers, although he obtained medals at all the great International Exhibitions. Notwithstanding that beautiful leather was thus shown to be made from them, the skins, when exposed to public auction in 1855, fetched only 2*s.* 7*d.* per half skin.

Since that period half-skins have reached the price of £2 5*s.* 6*d.*, and the leather has been often sold at 15*s.* to 21*s.* per. lb.

The skin when tanned has a great range of utility, comparing favourably with the best French kid in beauty, cheapness, and durability, whilst it is in high repute for the strongest shooting-boots.

A tanned porpoise skin and shoes made of it are in the Museum collection.

POLAR BEAR AND WHITE WHALE.

No individual has done more to develop the trade and increase their utility than Mr. G. Roberts of the Hudson's Bay Company. That company received the first porpoise skins from their settlement in Little Whale River, on the east coast of Hudson's Bay. They were imported as now in half skins in a salted state. Mr. Roberts undertook their development as leather on behalf of the company, and called with samples of the skins, and the leather that had been made in Canada from them, on several of the tanners in Bermondsey, who were all incredulous as to leather of such fine quality being produced from this skin. Ultimately he met a M. Bossard, a French tanner in business in Bermondsey, who took the matter up and produced from split skins, leather which up to this time has not been surpassed.

Although occasionally seen in the Bay of Chaleur and parts of New Brunswick, it is chiefly in the river St. Lawrence that this mammal is found. As early as 1707 there were no less than eight companies or bodies of men, established at different points of the river for capturing them. In the year 1710, one of these companies of six individuals took 800 porpoises: 159 porpoises were captured, in the years 1857 and 1858, at Rivière Ouelle. They are taken in enclosures made of light and flexible poles fixed in the beach, and harpooned or speared. The animal ranges in weight from 2,500 lbs. to 4,000 lbs, and each is probably worth for its oil and its skin £7 or £8. About 250 to 500 of these whales must be captured within the Hudson's Bay territories, for they have occasionally imported, in a year, 2,000 half skins.

Its flesh, which is said to be very good, is eaten by the inhabitants of some of the northern coasts where it is caught. Hans Egede says, "His flesh as well as the fat, has no bad taste, and when it is marinated with vinegar and salt, it is as well tasted as any pork whatever;" the fins also and the tail pickled or soused, are very good eating. Some of the internal membranes are used for windows, and some as bed-curtains; the sinews furnish the best sort of strong thread.

The *Delphinapterus leucas*, Pall., yields about 32 lbs. of oil: in some of the Russian seas the weight of blubber obtained is from 600 to 1000 lbs.

CHAPTER XIV.

MARSUPIALS AND THEIR USES.

The Wombat—Opossums, Australian and American—Various species of Kangaroos, Wallabies and Wallaroos, hunted for their skins and flesh—Kangaroo-leather—Fur cloaks—Smaller marsupials—Bandicoots—Koala or native cat, kangaroo rat, &c.

AUSTRALIA is distinguished by a large number of pouched animals, which are now extinct in almost every other part of the world; except in America, where a few species still exist. They are considered to be the oldest mammals known.

At a rough estimate there are 110 marsupials known in Australia. The existing species are of moderate growth, for the largest do not exceed two hundred pounds in weight; they are divided into carnivorous or flesh-eating, and herbivorous or grass-eating sections, with a few genera of mixed feeders.

THE WOMBAT (*Phascolomys Wombat*), of which there are several varieties in Australia, sometimes called the native pig, is a burrowing animal, feeding chiefly upon roots. It is stout and heavy-bodied, and is generally about three feet in length. Its fur is very coarse and its skin tough. Its flesh, although red and coarse in appearance, resembles that of a pig in flavour, and is usually cooked by the colonists as they would cook fresh pork. In some places the wombats are very numerous, and as they burrow near the surface, the blacks easily find the course of their underground passage by striking the earth and listening to the sound. By this means they trace out the animal, which they forthwith unearth and knock on the head, and, as is usual with them, roast it. The wombat is easily taken alive and domesticated.

OPOSSUMS.—The family of the *Carpophaga* (the *Phalanger*

tribe), has its stronghold in the southern portions of Australia, in particular on the south-east coast.

The common opossum (*Phalangista vulpina*) has the widest range, and is found in almost every part of Australia. The dark-coloured, ring-tailed Phalanger (*P. viverrina*) is found on the southern and western coast, and on the plains of the interior. A species of *Cuscus* inhabits the extreme north, where the handsome little *Belideus ariel* is also found. *Belideus sciurus* extends from the southern portions of Queensland to Rockhampton and Port Denison. The great flying phalanger (*Petaurista Taguanoides*) is shot only for his skin.

The opossums, of which there are several varieties, are very numerous in the open forest and brush lands; their habitation is the hollows of large trees, from which they rarely venture out except at night. Shooting opossums is a very favourite sport for moonlight nights, when a great many may be killed by one whose eye is quick at detecting them, and whose skill as a marksman is sufficiently good to bring them from their elevated perches. To a person who is well accustomed to its use, a long-barrelled pistol is better for this purpose than a gun, first, because it is a more portable article in scrambling through the brushwood, and secondly, because it is frequently necessary to fire at a point immediately above the head of the sportsman. The presence of the opossums may be detected by a peculiar shrill cry to which they give utterance, or by the marks made by their claws in ascending the trees. The opossum seldom weighs more than five or six pounds. His flesh is good, but is not much used by the settlers; it affords to the blacks, however, the largest portion of their animal food, as they are able to get the opossum with comparative ease, by climbing the tree on which it abides. A good dog will scent an opossum most readily; the more readily indeed, because he generally gets the carcase of the little animal as his share of the spoil, the sportsman contenting himself with the skin.

When the bushman's dog points out the *locale* of an opossum, the dog's master will often amuse himself by storming the dwelling of the concealed animal and capturing him at once; if he is in the hollow of a limb, the limb is cut off, and he is speedily turned out; if he is in the body of a hollow tree the interior of which is accessible near the root, a fire is lit below, and the poor

a, The Woolly Kangaroo (*Macropus laniger*), with young protruding from the "Marsupium." *b*, Opossum (*Didelphys ornata*).

opossum is compelled either to come out and face his enemies or to be suffocated—he chooses the former alternative, and is shot.

The Virginian opossum (*Didelphys virginiana*), one of the largest and most robust of the genus, is eaten in the Southern States of America, and is said to be white and well tasted, but its ugly tail is enough to put one out of conceit of the fare.

The Gamba, or saraque, another of the species, is considered a delicate meat by some in Brazil, while by others it is regarded with repugnance; the axillary glands have to be extracted before it can be eaten.

About twenty years ago opossum skins first came into demand to be made into furs for ladies, and large quantities were brought into Buffalo city from Pennsylvania and Ohio, selling at about *2s.* a-piece. In the Museum Collection of furs there are skins of the Australian and American opossums and of the kangaroo.

KANGAROOS.—There is no family of marsupials so largely represented and so widely distributed as the *Poephaga*, or kangaroos. They are found from Tasmania to New Guinea, and from the shores of the Pacific to those of the Indian Ocean. They are more or less nocturnal in their habits, shy and timid, and some as fleet as the wind. The female is provided with a pouch containing four teats, and produces one young at a time, though now and then two have been taken from the same pouch. The larger species are represented by the common great grey kangaroo (*Macropus major*) on the south and east side of Australia, and in the interior, by *Macropus ocydromus*, on the west coast, whilst another large kangaroo, *Osphranta antelopinus*, is found in the far north. The fine red kangaroo (*Osphranta rufus*) inhabits the plains of the interior, almost from east to west, never approaching the coast; while *Osphranta robustus* is only found in the mountainous district of the eastern parts.

Mr. George Masters states that a very large and savage species of kangaroo is said to inhabit some of the mountain ranges about Spencer's Gulf, in South Australia. There are also reports from various travellers that large kangaroos, far exceeding in size any known species, exist to the north in the interior of the country. At present, however, no specimens of these large creatures have come to hand. The smaller species of the genus *Halmaturus* are more restricted in their habitat; still, some roam over a very large extent of country. Peculiar to the east coast is the black wallaby (*Halmaturus ualabatus*), the only species found near Sydney. Further north occur *H. ruficollis*, principally on the

Clarence River, and on the borders of Queensland, another handsome wallaby, *H. dorsalis*, makes its appearance. *H. parryii* is also found in the northern districts of New South Wales. *H. parma* inhabits the semi-tropical Illawarra district, as does also the Pademelon wallaby (*H. thetidis*). In South Australia, we find *H. Greyi*, and *H. Derbianus*. On the west coast, *H. hautmanni*, *H. parma*, *H. brachyurus*, and the fleet and handsome *H. manicatus*. In Tasmania, *H. Billardieri*, and *H. Bennetii* occur.

Each part of the Australian coast has also one or more different kinds of *Petrogale* or rock wallaby, namely:—*P. penicillata*, *P. longicauda*, and *P. inornata* in New South Wales, *P. xanthopus* in South Australia, *P. lateralis* in West Australia, and *P. brachyurus* and *P. concinna* on the north-west coast.

The genus *Onychogalea*, or nail-tailed kangaroo, has a similar distribution. *O. frænata* is peculiar to the plains of the interior, and found from the eastern slopes of the coast range, to the junction of the Darling River, where *O. lunata* first appears, and extends within a short distance of the west coast. Another member of this genus, *O. unguifer*, inhabits the north-eastern portion of the continent. Many varieties (they can hardly be called species) of the genus *Lagorchestes*, or hare kangaroo, roam over the vast plains of the interior, where also the burrowing *Bettongias* or Jerboa kangaroos are principally found; the genus *Hypsiprymnus*, or rat kangaroos, inhabit the forests and scrubs.

There are a great many species of kangaroo, but the variation is not so much in appearance as in size and in their places of resort. The forester is the largest of the family, and is frequently found of two hundred pounds weight. The large males of this species are generally called the "old man kangaroo" by the colonists, and by such of the aborigines as have a smattering of English; an accomplishment, by-the-by, which few of them are without. The wallaby and the pademelon are much smaller, the former inhabiting rocky grounds, and the latter being found exclusively in what is termed brush lands. The average weight of the wallaby is about twelve or fourteen pounds, and that of the pademelon about nine or ten pounds. The kangaroo rat seldom weighs more than three or four pounds, and is found in various localities, even in the most barren scrub. It is not, as its name would import, anything of the rat species; but a perfect kangaroo in miniature. The rock wallaby is found in large numbers on many of the small islands near the coast, as well as the main land. One species, the Jerboa, or Kangaroo Mouse, discovered by Sir Thomas Mitchell, is not larger than the common field Mouse.

Chœropus occidentalis are very good eating (Krefft). The flesh of the short-nosed bandicoot (*Perameles obesula*) is delicious, especially when done in the native style, that is, the hair removed, and the game roasted upon the coals. The flesh of the great red kangaroo (*Osphranta rufus*) is very palatable, and preferable to that of *Macropus major*. The flesh of the nail-tailed kangaroo (*Onychogalea lunata*) is white and well tasted. That of the hare kangaroo (*Lagorchestes leporoides*) is delicious. That of the kangaroo rat (*Bettongia rufescens*) is also very palatable. It is less sought for than his larger relatives, except by the sawyers and splitters, to whom these animals yield many a fresh meal during their sojourn amidst the heavily-timbered flats and ranges. They are very numerous, but are seldom seen in the day time. At night, however, with the exercise of due caution they are easily shot. The plan which is found most successful is the following. Strew ashes or white sand over a small space of ground, so as to form a light-coloured spot upon which the dark forms of the kangaroo rats may stand out in relief amidst the surrounding darkness. At some distance—as far as possible, consistent with the clear discernment of the animal, and with the certainty of the shot—the sportsman must crouch down, having his gun pointed towards the white spot, and resting on a log or a forked stick. A little maize or some broken bread strewed over this white spot will attract the animals to it, and the hidden sportsman can hardly fail to secure one or more at a single shot. His movements, however, must be very slow and silent, and if he is concealed by a bush or a fallen tree, so much the better, for these little animals are very wary.

The flesh of all the kangaroos is good. The fore quarters, indeed, of the larger animals are somewhat inferior, and are usually given to the dogs; but from the hinder quarters some fine steaks may be cut. These, cooked in the same manner as venison collops, are, to most palates, very little inferior to the latter. The flesh of the large kangaroo, as well as that of the wallaby, is often dressed in the shape of a hash, and in this form, also, it is excellent. Mr. Gould asserts the excellence of kangaroo venison as a meat for the table. But the most admired part of the kangaroo is his tail. This is of enormous size in proportion to the rest of the body, the tail of a full-grown forester usually weighing ten or twelve pounds. It makes a superb soup, very much superior to ox-tail. The wallaby too is most commonly used for soup. The pademelon when cooked like hare affords a dish with which the most fastidious gourmand might be satisfied; the flesh of the kangaroo rat resembles that of a rabbit, and it eats best when cooked in the same manner; the last-mentioned animal is but little eaten, except by thorough bushmen, owing to the prejudice excited by the unfortunate name which has been bestowed upon it, but those who have once conquered this prejudice usually become fond of it.

Mrs. Millett, in her work "An Australian Parsonage," states that kangaroo venison still retains an honoured place at all tables in the interior. "The usual price of the meat, whilst in season, is 2½*d.* the pound, and the hind quarters and tail alone are cooked, the other parts of the animal falling to the share of the dogs. Being very dry meat it requires as much basting as a hare, and is generally eaten with the accompaniment of fat pork. A larded oin of kangaroo is a real dainty. The tail makes a splendid soup, which is very nutritious, and also capital stews and pasties: in fact, it is far superior to ox-tail for all culinary purposes. If kangaroo is dressed like jugged hare the deception is complete; and the plan also answers well of salting a piece of the loin and hanging it up the chimney to dry in the smoke until it becomes hard enough to be grated like Hamburg beef."

From the hind quarters of a huge forester, an animal which often weighs 200 lbs., some capital steaks may be cut, which, when cooked in the same manner, are very little inferior to venison collops.

The taste of the flesh of the full grown animal may be compared to lean beef, and that of the young to veal. They are destitute of fat, if we except a little occasionally seen between the muscles and integuments of the tail. The colonial dish, called a *steamer*, consists of kangaroo meat dressed with slices of ham. The liver, when cooked, is dry and crisp, and is considered a substitute for bread.

The larger kangaroos, which inhabit the open forest lands and the plains, are usually taken by coursing; a kind of sport in which, owing to the nature of the country and of the animal pursued, the horses, dogs, and huntsmen have generally their work to do. The kangaroo bounds off at a pace which requires his pursuers to use their best speed in following him, and the great height to which he can spring, enables him to clear all ordinary impediments to his progress. For the most part, therefore, his course is straightforward, through thick and thin, and in anything of a difficult country, more especially in some of the mountain regions, none but a good steed and a good rider can follow with any chance of success; the stock-horses bred in the colony are admirably adapted to this kind of service. Kangaroo hounds are also bred and trained in the colony by the lovers of this sport, and a good deal of training, indeed, is necessary; for the kangaroo, besides the speed and the peculiarity of his course, is a much more formidable antagonist to the canine race, and even to their biped masters, than would at first sight appear. The pademelon is sought for by the sportsman, either alone or aided by a good dog, and is brought down by the gun. A good deal of experience in bush-craft, however, is necessary for the acquisition of sufficient skill and quickness in this kind of shooting, owing to the rapidity with which the animal, on being alarmed, bounds away through the underwood to the concealment which the underwood affords him, and to the irregularity of his course. Old bushmen, nevertheless, are very successful in this sport, when they think fit to pursue it, and the blacks are peculiarly skilful in capturing and killing the pademelon after their own fashion.

"Boomers" (forester kangaroos) sometimes stand seven feet high. The average weight of a bush kangaroo is about eighteen to thirty pounds; the wallaby is still less. The latter are taken in large numbers, by means of wire snares. Their skins, when dried in the sun, fetch from one shilling to one shilling and sixpence each. Great numbers of kangaroos are slaughtered every year for their skins alone, and the hunters make good wages. They increase so fast when unmolested, as to interfere with the profits of the grazier. Kangaroo skins have become an important article of commerce. They make the toughest and most pliable of leather for boot uppers, the best of morocco, and are used for whips, gloves, &c. So far back as 1849, 12,500 kangaroo skins were exported from Western Australia. An export duty of one shilling is levied on them. 40,590 were imported into Melbourne in 1871, from the other colonies. Kangaroo skins (green) sell in South Australia at from 7*s.* to 45*s.* the dozen, and wallaby skins at 10*s.* to 25*s.* the dozen. Mr. James Farrell, of Melbourne, was awarded a premium of £100 by the Victoria Government some ten years ago for the industrial application of kangaroo skins, and they have now become an important article of traffic, for experts declare that they make the toughest and most pliable leather in the world. Boot uppers of this material are said to be both comfortable and durable. Of these skins some are exported in their raw state, and others are locally dressed and manufactured. 7,000 kangaroo skins were lately imported into San Francisco for tanning.

As showing the present great demand for skins, occasioned by the progress of furriery in Australia and in England—to which country there has of late sprung up and largely developed an exportation of pelts in the rough state—it may be mentioned that during the past five years the skins of the native Australian animals have advanced in value about sevenfold, so that kangaroo skins, which could formerly be bought for from 3*s.* 6*d.* to 8*s.* per dozen, are now sold at from 20*s.* to 60*s.*, and the coatings of wallabies now fetch from 36*s.* to 48*s.* a dozen instead of 48 pence. And it is the same with regard to opossum and nearly all other skins. Furriery, of course, has not alone caused this advance. There are now hundreds of thousands of native pelts exported to the English leather-merchants and furriers. One leading firm in Melbourne sends about 40,000 or 50,000 dozen away every season, and a London merchant, with whom they trade, recently offered to take a million skins at a price a trifle below their then marketable value. These facts will give some idea as to the vast number of skins annually sent from Australia.

The natives of Australia make cloaks of the skin of the kangaroo. It requires three kangaroo skins to make a large, full cloak, such as one of those worn by the women ; and the skins of the female kangaroo are preferred, those of the males being considered too thick and heavy. The skins are prepared by first stretching them out, and pegging them down on the ground in the shade. The women then with a native knife scrape off all the soft inner parts, and afterwards rub them well to soften them with grease and wilgi, or clay. To form the cloak, the skins are sewn together with the dried sinews of the tail of the kangaroo, or when these are not at hand, with a kind of rush (*Thysanotus fimbriatus*). The cloak is worn with the hairy side inwards.

The "Goulburn Herald" mentions the result of an annual battue made in 1876, by a party who were out for nine days in the mountains with guns on the Boorombi ranges on the Murrumbidgee, and brought home the skins of 1806 wallabies (*Halmaturus sp.*), and 23 wallaroos (*Osphranta robustus*).

Wallaby skins are very plentiful further north. It was stated in the Queensland Catalogue of the Vienna Exhibition, that there would be no difficulty in procuring a supply of 2,000 per week; "waxed," they are worth, in the colony, like the black grained kangaroo skins, 4*s.* 6*d.* per lb.; dressed with the fur on, 2*s.* 6*d.* each.

Of the Bandicoots, there are several species. In South Australia there are two found, *Perameles fasciata*, and *P. obesula*. The flesh of the rabbit-eared species (*P. lagotis*) is sweet, and resembles that of the rabbit, by which name it is known in Western Australia. Most of these Australian animals are clothed with rich and beautiful fur, and the conversion of this material into articles of dress and ornament affords profitable employment. A well-finished carriage rug of the Tasmanian black opossum is an article of luxury that it would be difficult to surpass, whether in usefulness or elegance. The animal known as the koala or native bear (*Phascolarctos cinereus*), the native cat (*Dasyurus*), the wallaby (*Halmaturus*), kangaroo rat (*Hypsiprymnus*), and some others, all supply the Australian furrier with valuable material, which he converts into men's caps, coats, waistcoats, &c., and into ladies' muffs, cuffs, and other articles of wear. The largest of the brush-tails (*Phascogale penicillata*) is found in almost every part of Australia, near the coast districts at least.

Fur rugs, of beautiful softness, close and warm, and often of elegant appearance, are annually made in thousands of the skins of the opossum (*Phalangista vulpina*) and the native cat (*Dasyurus viverrinus*), not only as carriage wraps but for use instead of blankets by the great number of people whose business leads them to sleep in the open air.

A good waterproof rug of sixty skins can be had in Australia for £2 or £3. Considerable purchases of opossum skins have of late years been made to send to England for the manufacture of gloves. Liberal prices have been offered by shippers, and it is hoped that this may induce some of the unemployed to turn their attention to procuring them. The opossum has for many years past been a great nuisance in the wooded country and hilly districts throughout the colonies. The white gum trees in many localities have been almost destroyed by them, and the injury they do to gardens, orchards, vineyards, and wheat crops is astonishing.

The Australian trapper carries on profitably in several parts of the country his lonely occupation ; and several hundreds of men pass six months of the winter season in capturing, for the sake of their furs, the opossum, native cat, and flying squirrel. The principal scene of the labours of these adventurers is the immense expanse of open woodland country that borders the arid plains of the Adelaide desert, and after passing Horsham, on the Wimmera, the solitary tents of these wandering sons of Nimrod may be seen at intervals of several miles. The method generally used to trap the unsuspecting varmint is by the old figure of four trap, a heavy weight being attached to the upper covering, which, on losing the support of the treacherous piece of wood, previously baited, falls heavily on the unfortunate animal, killing it instantly ; the skins by this method are uninjured. The trapper has generally each day 40 or 50 traps to attend to, leaving them baited every evening, and going round the following morning to gather the trophies of the chase ; of course he is compelled to be satisfied if only half of his traps have proved successful. The animals are then skinned, and the skins pegged out with the fur inside to the smooth bark of a gum tree, for the purpose of drying, and afterwards are packed away in bundles to be forwarded at the end of the season to Melbourne, where, we understand, good skins will fetch from 3*d.* to 4*d.* each, a successful trapper being often able, after paying all expenses, to clear £40 or £50 by his trip.

INDEX.

THE END.

BRADBURY, AGNEW, & CO., PRINTERS, WHITEFRIARS.

www.ingramcontent.com/pod-product-compliance
Lightning Source LLC
Chambersburg PA
CBHW021345210326
41599CB00011B/759

* 9 7 8 3 3 3 7 2 4 0 0 2 8 *